Fundamenta

Fundamenta

Fundamentals of Geology

Second Edition

Carla W. Montgomery
Northern Illinois University

WCB **Wm. C. Brown Publishers**
Dubuque, Iowa•Melbourne, Australia•Oxford, England

Book Team

Editor *Craig S. Marty*
Developmental Editor *Robert Fenchel*
Production Editor *Renee Menne*
Designer *David C. Lansdon*
Art Editor *Carla Goldhammer*
Photo Editor *Shirley Charley*
Permissions Editor *Gail I. Wheatley*
Art Processor *Brenda A. Ernzen*

Wm. C. Brown Publishers
A Division of Wm. C. Brown Communications, Inc.

Vice President and General Manager *Beverly Kolz*
National Sales Manager *Vincent R. Di Blasi*
Assistant Vice President, Editor-in-Chief *Edward G. Jaffe*
Director of Marketing *John W. Calhoun*
Advertising Manager *Amy Schmitz*
Director of Production *Colleen A. Yonda*
Manager of Visuals and Design *Faye M. Schilling*

Design Manager *Jac Tilton*
Art Manager *Janice Roerig*
Publishing Services Manager *Karen J. Slaght*
Permissions/Records Manager *Connie Allendorf*

Wm. C. Brown Communications, Inc.

Chairman Emeritus *Wm. C. Brown*
Chairman and Chief Executive Officer *Mark C. Falb*
President and Chief Operating Officer *G. Franklin Lewis*
Corporate Vice President, President of WCB Manufacturing *Roger Meyer*

Cover photo: Cirque of Towers, part of the Wind River Mountain
Range, in Bridger Wilderness in central Wyoming. © David Muench
Photography

Copyedited by *Carol Kozlik*

Copyright © 1989, 1993 by Wm. C. Brown Communications, Inc. All
rights reserved

Library of Congress Catalog Card Number: 92–85593

ISBN 0–697–09806–0

Unless otherwise indicated, all photographs © Carla W. Montgomery.

Printed in the United States of America by Wm. C. Brown Communications, Inc.,
2460 Kerper Boulevard, Dubuque, IA 52001

10 9 8 7 6 5 4 3 2 1

Contents

CHAPTER 18

Wind and Deserts 259

CHAPTER 19

Mass Movement, Mass Wasting 271

CHAPTER 20

Mineral and Energy Resources 285

List of Boxes

Preface

A professor once observed that everyone should take one course each in geology, botany, and astronomy, for everywhere one goes there are rocks, plants, and stars. Geology gives us a greater understanding of the planet upon which we live. As humans have come increasingly to depend on geologic resources, knowledge of geology has been important in realizing the origins and the limitations of those resources. We have also come to recognize that we can anticipate—and therefore avoid—some of the disasters that are a natural consequence of certain geologic processes; in a few cases, we may even be able to modify those processes for our benefit.

In practice, most students who study introductory geology fall into one of two groups. Some are prospective geology majors, for whom the wide perspective of physical geology is commonly the foundation on which more advanced coursework builds. A larger number are nonmajors, and indeed not prospective scientists of any kind, who are prompted to take the course by some mixture of interest in the subject and the need to satisfy a science distribution requirement. Although the needs of these two groups are not identical, every effort has been made to devise a text with sufficient versatility and appropriate learning aids to serve both kinds of students.

About the Book

This volume is, by and large, condensed and somewhat modified from my *Physical Geology* text. It attempts particularly to respond to the concern that modern geology is so broad in scope that many courses, especially those taught in one quarter, cannot cover all that is included in the standard texts. The principal omission from this text is the last chapter of the *Physical Geology* text, which deals with planetary geology. Most of the remaining chapters are streamlined by the elimination of material tangential to the main focus and some reduction in the number of terms introduced. However, many of the practical or environmental aspects and case histories have been retained, in consideration of their very positive reception by reviewers and users of the text.

This book is intended for an introductory-level college course in geology. It assumes no prior exposure to college-level mathematics or science. For the most part, metric units are used throughout, except where other units are conventional within a discipline. For the convenience of students not yet comfortable with the metric system, a unit conversion table is printed on the inside front cover of the book, and English-unit equivalents are occasionally given within the text also.

In the early chapters, the student is introduced to the nature of geology as a science; to the broad outline of the history of the earth by way of historical perspective; and to minerals and rocks, the building blocks of geologic features. Along the way, chapters on closely associated processes are included: The chapter on volcanic activity follows that on igneous rocks, and the chapter on weathering, erosion, and soil precedes the sedimentary-rock chapter. The set of chapters on such basic materials and processes is followed by a discussion of time in geology, how we measure it, and how it relates to our understanding of the earth.

Once this groundwork has been laid, the next several chapters explore the major features and processes of the earth's interior. This begins with plate tectonics, which provides the conceptual framework for understanding much about seismicity and volcanism. The chapter on earthquakes follows. The data of seismology and volcanology are the major source of information about the earth's interior. Moving from the interior toward the surface, the next two chapters deal, respectively, with the continental crust, including crustal structures and mountain building, and the ocean basins. Superimposed on these large-scale features are the effects of the surface pro-cesses, the subjects of the next several chapters. These include the various ways in which water, ice, wind, and gravity act to modify the surface features and forms of the earth. It is at the surface that we see the interaction between the internal heat that drives the internal processes and the solar energy that drives the winds and the water cycle. The final chapter is a brief survey of mineral and energy resources—what they are, how they form, aspects of supply/demand issues, and environmental impacts of their use. The appendix provides an identification chart for common minerals, and some guidelines for recognizing different rock types.

To the Instructor

The text organization just described places internal processes ahead of surface processes, and plate tectonics before the midpoint of the text. This puts discussion of the large-scale processes ahead of that of the more localized surface processes that act on the resulting features. The far-reaching concepts of plate tectonics are introduced in some detail as soon as the students have a background in earth materials and time on which to draw for understanding. This allows plate-tectonic concepts to be reinforced by references in subsequent chapters, as relevant to the topics under discussion. There also seems to be a certain logic in building features by tectonic processes before tearing them down by surface processes. However, the various chapters are relatively independent, so that an instructor who prefers to do so can cover surface processes ahead of internal processes with minimal difficulty. The glossary should help to bridge any gaps arising from reordering of blocks of the text.

A discussion of environmental geology has been added to some texts as a separate chapter. However, there are environmental aspects to many of the topics in the text, from volcanoes to streams to resources. Therefore, in this book, environmental and human-impact considerations are woven into various chapters as appropriate. This may help students to see the present relevance of the subject matter while mastering the corresponding facts, theories, and vocabulary.

A variety of pedagogical aids and features are included. Each chapter begins with an outline of the subject headings to follow, by way of overview. Terms are printed in boldface and defined at first encounter; these boldfaced terms are collected as "Terms to Remember" at the end of each chapter and are defined in the glossary for quick reference. At the end of each chapter are both questions for review to help students' study efforts and a small number of questions or problems for further thought that go beyond basic review of text material. There are also suggestions for further reading that include several kinds of material: up-to-date (but often relatively sophisticated) references in the subject area of the chapter, materials that may be more readable for the nonspecialist (including some older but fundamentally accurate works), and, occasionally, "classic" works by prominent geologists.

Most chapters contain one or more boxed inserts. These are of several types. Some describe tools of the geologic trade (for example, thin sections). Some present case studies related to chapter material (flood recurrence-interval projection for a particular stream, groundwater depletion in the Ogallala aquifer system, shoreline stabilization efforts along the Texas shore). A few present supplementary material related to the body of the text (the periodic table). In all cases, the material is included for enrichment or information without disrupting the flow of the main body of the text and the presentation of fundamental concepts. Individual boxes may be included or omitted at the instructor's discretion. Occasional "miniboxes" are also set off within the text. These could be viewed somewhat as long, parenthetical remarks, minor digressions not lengthy enough to justify a major boxed insert, and usually lacking associated figures.

Those who are familiar with the first edition of *Fundamentals of Geology* may be interested to know in what ways this second edition differs from the first. Some chapters have been resequenced, at the suggestion of adopters. For example, the chapter on volcanic rocks and processes now follows immediately after the chapter on igneous rocks, and the chapter on geologic time has been moved up to follow the rock chapters, just ahead of plate tectonics, so key concepts in geologic time and process rates are explored much earlier. Plate tectonics is introduced briefly in a boxed insert in the first chapter, too, so immediate reference can be made to aspects of it. Also at users' suggestion, the historical-geology chapter has been replaced by one on geologic resources; though the subject of resources falls somewhat outside the traditional focus of physical geology, it is a subject of undeniable importance and current interest to both students and instructors. The art program has been substantially improved, both through refinements in the line art and through replacement of a number of photographs with better and/or clearer examples. The text has been updated in several respects: in terms of the technical material (for instance, seismic tomography is now incorporated into the chapter on the earth's interior); in the inclusion of current data, as in the resource chapter, and newer references in many chapters; and in incorporation of recent geologic events, such as the Loma Prieta earthquake, *Exxon Valdez* spill, or ongoing volcanic activity in the Aleutians and Hawaii.

Acknowledgments

A great many individuals have contributed to this project, directly or indirectly. Much of the inspiration for it has come from the many beginning geology students I have taught, who continually remind me of the fascination of discovering geology for the first time and motivate me to find new ways to share that excitement. The push to move from inspiration to reality has, in turn, come from the enthusiastic (and persistent!) folks at WCB.

I would like to renew my thanks to the reviewers of the first edition of *Fundamentals of Geology:* E. Hasenohr, Eastern New

Mexico University; R. Schiffman, Bakersfield College; N. Fields, Western Kentucky University; K. Koenig, Texas A&M University—Main; L. Crow, Houston Community College; L. Cummins, Angelo State University; V. Scott, Oklahoma State University; L. Boyer, University of Wisconsin—Milwaukee; and J. Stewart, Vincennes University. I would also like to acknowledge the assistance and direction provided from the outset by those additional instructors who responded to a text development survey: J. Lyman, Bakersfield College; M. Bikerman, University of Pittsburgh; G. Jacobson, Grossmont College; H. Level, Ventura College; R. Krauth, Middlesex County College; L. McClure, Bloomsburg University; J. Ratzlaff, Fort Hays State University; J. Droste, Indiana University at Bloomington; S. Gaby, Mohegan Community College; K. Kuehn, Western Kentucky University; and F. Goldstein, Trenton State College. The second edition of *Fundamentals* has in turn been improved and strengthened through the thoughtful suggestions and conscientious criticisms of reviewers Robert A. Schiffman, Bakersfield College; David M. Best, Northern Arizona University; L. Don Ringe, Central Washington University; Richard U. Birdseye, University of Southwestern Louisiana. The input from all of these colleagues is sincerely appreciated. Any remaining shortcomings in the text are, of course, my responsibility.

Thanks go to Carol Edwards and Joe McGregor at the U.S. Geological Survey Photographic Library for their continuing interest in and assistance with the photo research, and to fellow author and dean Jerrold H. Zar for his encouragement and understanding of my occasional absorption in textbook writing (and empathy with the more traumatic aspects of the process!). The quality of the finished book can be attributed in large measure to the book team at WCB. Continued association with these energetic, talented, and dedicated professionals has been a very rewarding (if occasionally exhausting) experience. Last, but assuredly not least, I would like to thank my husband, Warren, part-time photographer and field assistant, for the remarkable patience and supportiveness he has shown as yet another book-revision project loomed on our horizon.

C H A P T E R   1

Outline

An Invitation to Geology

The earth is constantly changing. These lavas,
new rocks extracted as melts from the interior,
are now being weathered away under the
combined attack of wind and water at the
surface. Keanae Point, Maui, Hawaii.
© David Muench

Introduction

Geology is the study of the earth and the processes that shape it. **Physical geology,** in particular, is concerned with the materials and physical features of the earth, changes in those features, and the processes that bring the changes about. Intellectual curiosity about the way the earth works is one reason for the study of geology. Piecing together the history of a mountain range or even a single rock can be exciting. Anywhere one goes, there are rocks and other earth materials present, and geologic processes at work. There are also very practical aspects to the study of geology. Certain geologic processes and events may be hazardous (figure 1.1), and a better understanding of such phenomena may help us to minimize the risks. We have also come to depend heavily on certain earth materials for energy or as raw materials for manufacturing (figure 1.2), and knowledge of how and where those resources are to be found can be very useful to modern society. Before entering into the detailed study of physical geology, we briefly survey geology as a discipline, and the history of the earth that is its subject.

Geology As a Discipline

In some sense, geology is a particularly broad-based discipline, for it draws on many other sciences. Knowledge of physics contributes to an understanding of rock structures and deformation, and supplies tools with which to investigate the earth's deep interior indirectly. The chemistry of geologic materials provides clues to their origins and history. Modern biological principles are important in studying ancient life forms. Mathematics provides a quantitative framework within which geologic processes can be described and analyzed. Physical geographers study the earth's surface features much as some geologists do. What makes geology a distinctive discipline is, in part, that it focuses all these approaches, and others, on the study of the earth. Moreover, having the earth as a sub-

Figure 1.1 Aftermath of the 1989 Loma Prieta earthquake: collapse of I-880.
Photograph by M. Rymer, from USGS Open-File Report 89–687

Figure 1.2 An open-pit iron mine: Empire Mine, Palmer, Michigan. Each of the steplike ledges is approximately 15 meters high.

ject introduces some special complexities not common to most other sciences.

The Issue of Time

The modern earth has been billions of years in the making. As will be seen in later chapters, many of the processes shaping the earth are extremely slow on a human time scale, some barely detectable even with sensitive instruments. It is therefore difficult to observe or to demonstrate directly, in detail, how certain earth materials or features have formed.

Furthermore, materials may respond differently to the same forces, depending on how those forces are applied. This can be seen, for example, in the phenomenon of fatigue of machine parts, in which a material quite strong enough to withstand a certain level of sustained stress fails during the application of a much smaller stress that has been repeated many times. As a prac-

Figure 1.3　Mount St. Helens in eruption, May 1980.
Photograph by P. W. Lipman, courtesy of U.S. Geological Survey

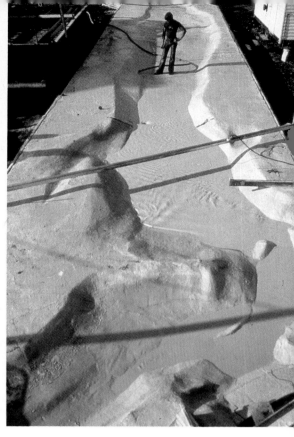

Figure 1.4　To study response of Swift Dam and Reservoir to mudflows from Mount St. Helens, a scale model is built.
Photograph by J. E. Peterson, USGS Photo Library, Denver, Colorado

tical matter, then, it may be impossible to duplicate some geologic processes in the laboratory because the human lifetime is simply too short.

The Matter of Scale

Likewise, some natural systems are just too large to duplicate in the laboratory. A single crystal or a small piece of volcanic rock can be studied in great detail. But it is hardly possible to build a volcano (figure 1.3), or a whole continent, in a laboratory to conduct experiments on them. Scale-model experiments, in which the materials and the forces applied to them are scaled down proportionately, are one compromise (figure 1.4). For example, in studies of designs for earthquake-resistant buildings, model structures are shaken by small vibrations in the hope that the results will mimic the response of large buildings to great earthquakes. Scale modeling is an

inexact science, however, and not all natural systems lend themselves to such studies.

Complexity in Natural Systems

A laboratory scientist about to conduct an experiment tries to minimize the number of variables, so as to obtain as clear a picture as possible of the effect of any one change—whether of temperature, pressure, or the quantity of some particular substance present. Typically, the experimental materials are kept simple: a single rock type or mineral, chemically quite pure. Natural geologic systems, however, are rarely so simple. Natural rocks and minerals invariably contain chemical impurities and physical imperfections; many different rocks and minerals may be mingled together; temperature and pressure may change simultaneously, while gases, water, or other chemicals flow in and out. Extrapolating from

carefully controlled laboratory experiments to the real world becomes correspondingly difficult.

Also, the laboratory scientist can perform an experiment in stages, examining the results after each step. The geologist, however, may be confronted by rocks or other materials or structures that have been altered or reworked several times, perhaps dozens of times, each time under different conditions, over millions or even billions of years. That history can be difficult to decipher from the present end product, particularly because the same end product can often be formed from several different possible starting materials, via different combinations of geologic processes, just as one can arrive at a given spot by traveling in various ways from various starting points.

Geology and the Scientific Method

The **scientific method** is a means of discovering basic scientific principles. One begins with a set of observations and/or a body of data, based on measurements of natural phenomena or on experiments. One or more **hypotheses** are formulated to explain the observations or data; there is also the possibility that no systematic relationship exists among the observations (*null hypothesis*). A hypothesis can take many forms, ranging from a general conceptual framework or model describing the functioning of a natural system, to a very precise mathematical formula relating several kinds of numerical data. What all hypotheses have in common is that they are unproven, and they must be susceptible to testing.

In the classical conception of the scientific method, one uses a hypothesis to make a set of predictions and then devises and conducts experiments to test each hypothesis to determine whether experimental results agree with predictions based on the hypothesis. If they do, the hypothesis gains credibility. If not, if the results are unexpected, the hypothesis must be modified to account for the new data. Several cycles of modifying and retesting of hypotheses may be required before a hypothesis that is consistent with all the observations and experiments that one can conceive is developed. A hypothesis that is repeatedly supported by new experiments advances in time to the status of a **theory,** a generally accepted explanation for a set of data or observations.

This approach is not strictly applicable to many geologic phenomena because of the difficulty of experimenting with natural systems. In such cases, hypotheses are often tested entirely through further observations and modified as necessary until they accommodate all the relevant observations. This broader conception of the scientific method is well illustrated by the development of the theory of plate tectonics, discussed in chapter 9. Even a well-accepted theory, however, may ultimately be found to require extensive modification. In the case of geology, a common cause of this is the development of new analytical or observational techniques, which make available wholly new kinds of data that were unknown at the time the original theory was formulated.

In addition to hypotheses and theories is a smaller body of scientific **laws:** fundamental, typically simple principles or formulas that are invariably found to be true. Included in this category are Newton's law of gravity; the principle of physical chemistry that states that heat always flows from a warmer body to a colder one, never the reverse; and the geologic law that states that, in an undisturbed sequence of layered sediments, the layers on the bottom must have been deposited before the overlying layers.

Key Concepts in the History of Geology

Humans have wondered about the earth in some way for thousands of years. The ancient Greeks measured it and recognized fossils preserved in its rocks as remains of ancient life forms. Theologians, philosophers, and scientists have speculated on its age for centuries. The systematic study of the earth that constitutes the science of geology, however, has existed as an organized discipline for only about 250 years. In its early years, it was predominantly a descriptive subject. Two principal opposing schools of thought emerged in the eighteenth and nineteenth centuries to explain geologic observations.

One, popularized by James Hutton and later named by Charles Lyell, was the concept of **uniformitarianism.** Sometimes condensed to the phrase, "The present is the key to the past," uniformitarianism comprises the ideas that the surface of the earth has been continuously and gradually changed and modified over the immense span of geologic time and that, by studying the geologic processes now active in shaping the earth, we can understand how it has evolved through time. It is not assumed that the *rates* of all processes have been the same throughout time but rather that the nature of the processes is similar—that the same physical principles operating on the earth in the past also are operating in the present. This point is explored further in the next section.

James Hutton was a remarkably versatile individual—physician, farmer, and only part-time geologist. In the early days of science, many advances in understanding were made by nonspecialists capable of careful observation and logical deduction. As various disciplines become more advanced and more sophisticated, however, the amateur is far less able to play a major role; too much accumulated knowledge must be assimilated to arrive at the forefront of research.

The second, contrasting theory was **catastrophism.** The catastrophists, led by Georges Cuvier, saw a series of immense, worldwide upheavals as the agents of change and assumed a static, unchanging earth between catastrophes. Violent volcanic eruptions followed by torrential rains and floods were invoked to explain mountains and valleys and to bury animal populations that later became fossilized. In between those episodic global devastations, the earth's surface did not change, according to catastrophist theory. Entire plant and animal populations were created anew after each such event, to be wholly destroyed by the next.

Is the Present the Key to the Past?

More detailed observations and calculations, greater use of the allied sciences, and the development of increasingly sophisticated instruments have collectively provided overwhelming evidence in support of the uniformitarian view. The great length of the earth's history makes it entirely plausible that processes that seem gradual and even insignificant on a human time scale could, over those long spans of time, create the modern earth in all its geologic complexity.

Figure 1.5   Meteor Crater, Arizona.

This is not to say that earth history has not been punctuated occasionally by sudden, violent events that have had a substantial impact on a regional or global scale. Geologists see evidence of past collisions of large meteorites with the earth (figure 1.5) and find volcanic debris from ancient eruptions that would dwarf that of Mount St. Helens. But these events are unusual and, for the most part, of only temporary significance in the context of the earth's long history. By and large, the modern earth is the product of uncounted small and gradual changes, repeated or continued over very long spans of time.

The same physical and chemical laws can be presumed to have operated throughout earth's history. Thus, by observing modern geologic processes, geologists can learn much about how those same processes might have shaped the earth in the past. The relative importance of each process, however, may not always have been just the same as it is now; nor can it be assumed that all processes operated in detail just as they do now. Certain irreversible changes in the earth have no doubt changed the nature and intensity of corresponding geologic processes. For example, the earth has been slowly losing heat ever since it formed. Present internal temperatures must be substantially lower than they were several billion years ago, especially near the surface. The earth's internal heat plays a key role in melting rocks and thus in volcanic activity. It is reasonable to infer that melting in the interior was more extensive in the past than it now is, that the products of that more extensive melting might be somewhat different from modern volcanic rocks, and also that volcanic activity was more extensive in the past. The earth's atmosphere, too, has undergone profound changes, as is explored further shortly. Briefly, the atmosphere has gone from oxygen-poor to oxygen-rich, and this change, in turn, necessarily affects the chemical details of such processes as weathering of rocks by interaction with air and water. However, geologists can determine, experimentally and theoretically, how rocks would react with the kind of atmosphere deduced for the early earth and thus characterize ancient weathering processes even though they cannot observe the exact equivalents in nature today. The earth may change; the physical laws do not. This concept is fundamental to modern uniformitarianism.

The Earth within the Universe

In recent decades, scientists have been able to construct an ever-clearer picture of the origins of the solar system and of the universe itself. From observations that the stars are all moving apart from each other came the recognition of an expanding universe. Clearly, that expansion cannot have been going on forever: If one extrapolates the stars' movements backward, a point is reached at which all matter was apparently together in one place.

Most astronomers now accept a cataclysmic explosion, or "big bang," as the origin of the modern universe. At that time, enormous quantities of matter were synthesized and flung violently apart across an ever-larger volume of space. The time of the so-called big bang can be estimated in several ways. Perhaps the most direct is the back-calculation of the universe's expansion to its apparent beginning by extrapolating the present motions of the stars backward in time until they converge. Various age estimates of the modern universe overlap in the range of 15 to 20 billion years.

The Early Solar System

The sun and its system of circling planets, including the earth, are believed to have formed from a single rotating cloud of gas and dust, starting nearly 5 billion years ago. Most of the mass coalesced at the center of the cloud, or nebula, to form what would eventually become the sun. Like the rest of the universe, the early sun consisted mostly of hydrogen, the principal product of the "big bang." The inner parts of this enormous ball of gas were so compressed that they became hot and dense enough to initiate nuclear reactions. The ball of gas became a star, radiating light and other forms of energy. Our sun is now a middle-aged star, having consumed about half of its nuclear fuel over the past 4.5 to 5 billion years. It should continue to shine for about 5 billion years more before it has used up so much of its fuel that it collapses to a cold dwarf and turns off the earth's solar energy.

While the proto-sun developed, the remaining matter settled into a rotating disk around it. Dust began to condense from the gas, and the dust gradually formed planets that continued to circle the sun as they formed (figure 1.6). Modern methods of dating rock material (see chapter 8) have shown the oldest fragments of meteorites and moon rocks to be close to 4.6 billion years old. Formation of the solar system is thus believed to have been substantially complete more than 4.5 billion years ago.

The Planets

The composition of each planet depended strongly on how near it was to the hot young sun. Very close to the sun, temperatures were so high that, at first, nothing solid could exist at all. As cooling progressed, the solids that condensed nearest the sun contained mainly high-temperature materials: metallic iron and a few other minerals with very high melting temperatures. The planets nearest to the sun, then, consist mostly of these materials. Somewhat farther out, where temperatures were lower, the developing planets incorporated much larger amounts of lower-temperature materials, including some that contain water locked within their crystal structures. (This eventually made it possible for the earth to have liquid water at its surface.) Still farther from the sun, temperatures were so low that nearly all of the materials in the original gas cloud condensed—even materials like

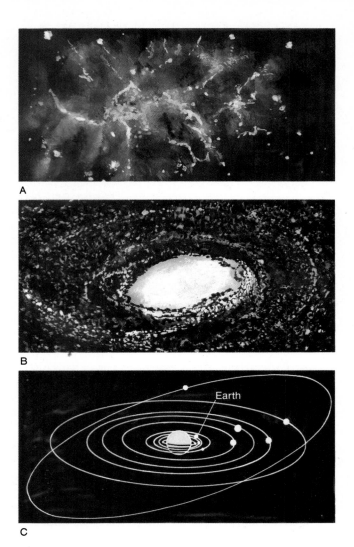

A

B

C

Figure 1.6 So-called nebular model of solar-system formation. (*A*) Beginnings as a rotating gas cloud. (*B*) Most of the mass becomes concentrated at the center to form the sun; remaining material condenses and accumulates to form planets. (*C*) Present solar system. Earth is the third planet from the sun. It is one of the smaller planets, but it has a unique combination of composition (including surface water) and climate that makes possible life as we know it.

methane and ammonia, which are gases at normal earth surface temperatures and pressures. Each planet, then, formed from an accumulation of bits of these condensed materials drawn together by gravity. Uncondensed gases were swept out of the interplanetary spaces by streams of matter and energy radiating from the young sun.

These solar-system-forming processes led to a series of planets with a variety of compositions, most quite different from that of earth. This is something to keep in mind when considering the possibility of someday mining other planets for needed minerals. Both the basic chemistry of these other bodies and the kinds of ore-forming or other resource-forming processes that might occur on them would differ considerably from those on earth and might not lead to products we would find useful. (This is leaving aside the economics or technical practicality of such mining activities!) In addition, principal current energy sources required living organisms to form, and so far, no life has been found on other planets or moons.

The Early Earth

The earth has changed considerably since its formation, undergoing some particularly profound changes in its early history.

Heating and Differentiation

Like most of the planets, the earth is believed to have begun as a sort of "dust ball" of small bits of condensed material collected together by gravity, with no free water, no atmosphere, and a very different surface from its present one. The dust ball was heated in several ways. The impact of the colliding particles as they came together to form the earth provided some heat. Much of this heat was radiated out into space, but some was trapped in the interior of the accumulating earth. As the dust ball grew, compression of the interior by gravity also heated it. (That materials heat up when compressed can be demonstrated by pumping up a bicycle tire and then feeling the barrel of the pump.) Furthermore, the earth contains small amounts of several naturally radioactive elements that decay, releasing energy (see chapter 8). These three heat sources combined to raise the earth's internal temperature enough that parts of it, perhaps eventually most of it, melted. The earth may at one time have had a layer of molten rock at its surface.

When accretion of material was complete, slow cooling set in. Metallic iron, being very dense, sank into the middle of the earth. As cooling progressed and remaining melt began to crystallize, lighter, low-density minerals floated out toward the surface. The eventual result was an earth differentiated into several major compositional zones: a large, iron-rich **core;** a thick surrounding mineral **mantle;** and a thin, low-density **crust** at the surface (see figure 1.7).

Chapter 11 examines how geologists know the structure and composition of the earth's interior. It can be shown that the earth's core is made up mostly of iron, with some nickel and a few minor elements, and that the mantle consists mainly of the elements

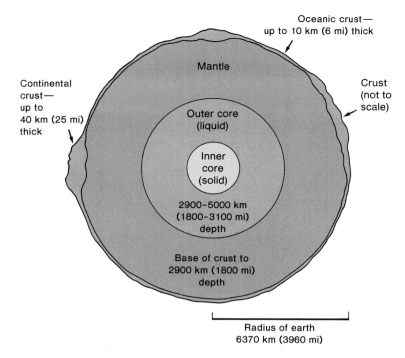

Figure 1.7   A chemically differentiated earth (crust not drawn to scale). Oceanic crust, which forms the sea floor, has a composition somewhat like that of the mantle but is richer in silicon. Continental crust is thicker and contains more low-density minerals rich in calcium, sodium, potassium, and aluminum. It rises above both the sea floor and the ocean surface.

iron, magnesium, silicon, and oxygen combined in varying proportions in several different minerals. The crust is much more varied in composition and very different chemically from the average composition of the earth, as is also discussed in chapter 11. The differentiation process was complete at least 4 billion years ago.

The Early Atmosphere and Oceans

The heating and subsequent differentiation of the early earth led to another important result: formation of the atmosphere and oceans. Many minerals that had contained water or gases locked in their crystals released them during the heating and melting. The early earth was much hotter than at present and subject to more extensive volcanic activity, with water among the gases thus released. As the earth's surface cooled, the water could condense to form the oceans. Without this abundant surface water, which in the solar system is unique to earth, most life as we know it could not exist.

The earth's early atmosphere was quite different from the modern one, even disregarding the effects of modern pollution. The first atmosphere had little or no free oxygen in it. Humans could not have survived in it. Oxygen-breathing life could not exist before the first simple plants—the single-celled blue-green algae—appeared in large numbers to modify the atmosphere. Their remains are found in rocks as old as several billion years. They manufacture food by photosynthesis, using sunlight for energy and releasing oxygen as a by-product. In time, enough oxygen accumulated that the atmosphere could support oxygen-breathing organisms.

Subsequent History

Differentiation produced profound internal changes in the earth and also established a combination of land, air, and water at the surface. Significant changes have continued at and near the surface through time.

BOX 1.1

Revolution in Earth's Evolution: Plate Tectonics

Rarely has a new geologic theory come so rapidly to dominate the discipline and, at the same time, to capture public attention. Plate tectonics, which became generally accepted only about two decades ago, now provides the conceptual framework that allows geologists to understand much about the nature of mountain building and other processes that shape our planet's surface.

Tectonics is the study of large-scale movement and deformation of the earth's crust. The basic concept of **plate tectonics** is simple: that the earth has an outermost rigid, rocky shell, consisting of the crust and uppermost mantle, broken up into a series of *plates* that can move on an underlying soft or plastic layer in the upper mantle (figure 1). Plates can collide or be split apart; as two plates converge, one can override the other. Continents shift in position on the globe as the plates of which they are a part move.

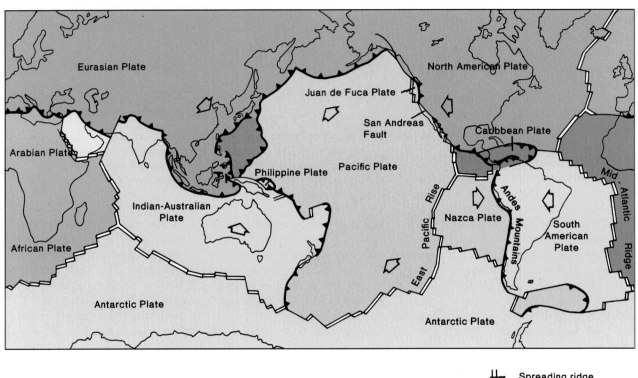

Figure 1   Principal world lithospheric plates.
Source: After W. Hamilton, U.S. Geological Survey.

The stresses associated with plate movements and deformation can cause earthquakes, especially at the boundaries between plates. Where one plate plunges below another into the hot mantle, melting occurs, and the melt may rise up into the plate above to erupt from a volcano (figure 2).

Over the long spans of geologic time, rocks are continually created, destroyed, and changed, in large measure as a consequence of plate-tectonic processes. These processes have been operating for billions of years.

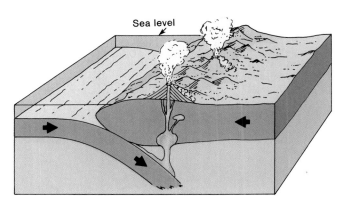

Figure 2 Ocean-continent convergence. The melting accompanying sinking of the down-going plate produces volcanic mountains on the continent.

The Changing Face of Earth

After the early differentiation, the earth's crust with its continents and ocean basins did not look the way we know it today. For one thing, the continents have moved (box 1.1). They have not always been the same size and shape, either. Rocks that were deposited in ocean basins can now be found high and dry on land, revealing the former presence of inland seas followed by great uplift. New pieces have been added to the edges of continents. Volcanoes once erupted where none now exist, leaving behind evidence of their earlier activity in ancient volcanic rocks. Tall mountains have been built up and then eroded away, sometimes several times in the same place, over billions of years.

Geologists can to some extent reconstruct the distribution of land, water, and surface features as they were at times in the past and identify geologically active areas such as developing mountain ranges, on the basis of the kinds of rocks or fossils of each age found and what is known of how such rocks formed or in what setting the fossilized creatures lived. Such reconstruction becomes more difficult the farther back geologists try to go in time, for the oldest rocks have often been covered by younger ones or disrupted by more recent geologic events.

Figure 1.8 The geologic spiral. Important plant and animal groups are shown where they first appear in significant numbers. All complex organisms—especially humans—have developed relatively recently in the geologic sense.
Source: *Geologic Time*, U.S. Geological Survey publication.

Life on Earth

The rock record shows when different plant and animal groups appeared. Some are represented schematically in figure 1.8. The earliest creatures left very limited remains because they had no hard skeletons, teeth, shells, or other hard parts that could be preserved in rocks. The photosynthetic microorganisms are known principally from remains of their communities—algal mats, for instance (figure 1.9). The first multicelled oxygen-breathing creatures probably developed about 1 billion years ago, after sufficient oxygen had accumulated in the atmosphere. However, they are poorly preserved, probably because they were entirely soft-bodied. By about 600 million years ago, marine animals with shells had become widespread.

Figure 1.9 These layered structures can be identified by a trained eye as *stromatolites*, fossilized algal mats, here found in rocks more than 2.5 billion years old.

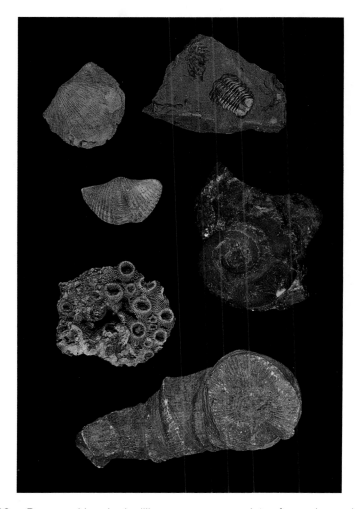

Figure 1.10 By several hundred million years ago, a variety of organisms with hard, preservable bones, shells, and other parts had become widespread.
© Wm. C. Brown Communications, Inc./Photograph by Bob Coyle

The development of organisms with preservable hard parts greatly increased the number of preserved animal remains in the rock record; consequently, biological developments since that time are far better understood than are those of earth's first several billion years (see figure 1.10). Dry land was still barren of large plants or animals half a billion years ago. In rocks about 400 million years old is the first evidence of animals with backbones— the fish—and also of some land plants. Insects appeared approximately 300 million years ago. Later, reptiles and amphibians moved onto the continents. The dinosaurs appeared about 200 million years ago and the first mammals at nearly the same time. Warm-blooded animals took to the air 150 million years ago with the devel-opment of birds, and by 100 million years ago, both birds and mammals were well established.

Such knowledge has practical applications. Certain of our energy sources have been formed from plant or animal remains. Knowing the times at which particular groups of organisms appeared and flourished is helpful in assessing the probable amounts of these energy sources available and in concentrating the search for these fuels in rocks of appropriate ages.

On a time scale of billions of years, human beings have just arrived. The most primitive human-type remains are no more than 3 to 4 million years old, and modern rational humans (*Homo sapiens*) developed only about half a million years ago. Half a million years may sound like a long time, and it is if compared to a single human lifetime. In a geologic sense, though, it is a very short time. Nevertheless, we humans have had an enormous impact on the earth, at least at its surface, an impact far out of proportion to the length of time we have occupied the planet.

The Modern Dynamic Earth

Although it has been cooling for billions of years, the earth still retains enough internal heat to drive large-scale mountain-building processes, to cause volcanic eruptions, to make continents mobile, and indirectly to trigger earthquakes. At the same time, the continual supply of solar energy to the earth's surface drives many of the surface processes: Water is evaporated from the oceans to descend again as rain and snow to feed the rivers and glaciers that sculpture the surface; differential heating of the surface leads to formation of warmer and colder air masses above it, which, in turn, produces atmospheric instability and wind.

Dynamic Equilibrium in Geologic Processes

Natural systems tend toward a balance or equilibrium among opposing forces. As new dissolved minerals are washed into the sea by rivers, sediments are deposited in the ocean basins, removing dissolved chemicals from solution. Internal forces push up a mountain; gravity, wind, water, and ice collectively act to tear it down again. When one factor changes, other compensating changes occur in response. If the disruption of a system is relatively small and temporary, the system may in time return to its original condition, and evidence of the disturbance is erased. A coastal storm may wash away beach vegetation and destroy colonies of marine organisms living in a tidal flat, but when the storm has passed, new organisms start to move back into the area, new grasses begin to establish roots in the dunes. The violent eruption of a volcano like Krakatoa or Mount St. Helens may spew ash high into the atmosphere, partially blocking sunlight and causing the earth to cool, but within a few years, the ash settles back to the ground and normal temperatures are restored.

This is not to say that permanent changes never occur in natural systems. The size of a river's channel depends, in part, on the maximum amount of water it normally carries. If long-term climatic or other conditions change so that the volume of water regularly reaching the stream increases, the larger quantity of water will, in time, carve out a correspondingly larger channel to accommodate it. The soil carried downhill by a landslide certainly does not begin moving back upslope after the landslide is over; the face of the land is irreversibly changed. Even so, a hillside forest uprooted and destroyed by the slide may within decades be replaced by fresh growth in the soil newly deposited at the bottom of the hill.

Figure 1.11 Subsidence over old underground coal mines, Sheridan County, Wyoming.
Photograph by C. R. Dunrud, USGS Photo Library, Denver, Colorado

The Impact of Human Activities; Earth As a Closed System

For all practical purposes, the earth is a **closed system,** meaning that the amount of matter in and on the earth is fixed. No new elements are being added. There is, therefore, an ultimate limit to how much of any metal we can exploit. There is also only so much land to live on. Conversely, any harmful elements we create remain in our geologic environment, unless we take some extraordinary step, such as expending the funds, materials, and energy required to cast something out into space via spacecraft.

Human activities can cause or accelerate permanent changes in natural systems. The impact of humans on the global environment is broadly proportional to the size of the population, as well as to the level of technological development achieved. This can be illus-

trated especially readily in the context of pollution. The smoke from one camp fire pollutes only the air in the immediate vicinity; by the time that smoke is dispersed through the atmosphere, its global impact is negligible. The collective smoke from a century of industrialized society, on the other hand, has caused measurable increases in several atmospheric pollutants worldwide, and those pollutants continue to pour into the air from many sources.

Likewise, in the context of resources, recent human impact has been dramatic. The world's population has soared from 2 billion to over 5 billion in little more than fifty years. That fact, combined with desires for ever-higher standards of living around the world, has created voracious demand for mineral and energy resources. Yet, although the population is growing, the earth is not. Therefore, our resources

are in finite supply, and some could be exhausted within our lifetimes. The very presence of humans, along with their building and farming activities, alters the landscape, and does so to an increasing extent as the population grows (figure 1.11). Superimposed on the natural geologic processes of change on the earth, then, are further changes, deliberate or unconscious, brought about by human activities. Many of the latter are occurring more rapidly than the compensating natural processes.

The principal focus of this text is natural geologic processes. However, from time to time, particularly significant human impacts on geologic systems, as well as the reverse—geologic impacts on human activities (such as the hazards associated with floods, earthquakes, and other geologic processes)—are highlighted.

Summary

Geology is the study of the earth. It is not an isolated discipline but draws on the principles of many other sciences. The earth is a challenging subject, for it is old, complex in composition and structure, and large in scale. Physical geology focuses particularly on the physical features of the earth and how they formed. Observations suggest that, for the most part, those features are the result of many individually small, gradual changes continuing over long periods of time, punctuated by occasional unusual, cataclysmic events.

The solar system formed from a cloud of gas and dust close to 5 billion years ago. The compositions of the resulting planets reflect their proximity to the evolving sun. Shortly after its formation, the earth underwent melting and compositional differentiation into core, mantle, and crust. The early atmosphere and oceans formed at the same time. Heat from within the earth and from the sun together drive many of the internal and surface processes that have shaped and modified the earth throughout its history and continue to do so. The earliest life forms date back several billion years; organisms with hard parts became widespread about 600 million years ago. Humans only appeared 3 to 4 million years ago, but their large and growing numbers and technological advances have resulted in significant impacts on natural systems, some of which may not readily be erased by slower-paced geologic processes.

Terms to Remember

catastrophism	mantle
closed system	physical geology
core	plate tectonics
crust	scientific method
hypothesis	theory
law	uniformitarianism

Questions for Review

1. What is *physical geology*?
2. Describe how the time factor complicates our attempts to understand geologic processes.
3. What is the scientific method? To what extent is it applicable to geology?
4. Compare and contrast the concepts of uniformitarianism and catastrophism.
5. How can the time of the "big bang" be determined?
6. Briefly summarize the process by which the solar system formed. Does this process lead to planets that are similar or quite different in composition? Explain.
7. What are the principal compositional zones of the earth?
8. How were the earth's atmosphere and oceans first formed? How and why has the atmosphere changed through time?
9. The earth is a closed system. Explain this concept in the context of resource use or of pollution.

For Further Thought

1. If the whole 4½-billion-year span of earth's history were represented by one twenty-four-hour day, how much time would correspond to (a) the 600 million years of existence of complex organisms with hard parts and (b) the half-million years that *Homo sapiens* has existed?
2. Many geologic processes proceed at rates on the order of one centimeter per year (1 inch = 2.54 centimeters). At that rate, how long would it take you to travel from home to school, or from your room to class?

Suggestions for Further Reading

Cameron, A. G. W. 1975. The origin and evolution of the solar system. *Scientific American* 233 (September): 32–41.

Cloud, P. 1988. *Oasis in space: Earth history from the beginning.* New York: W. W. Norton.

Eicher, D. L., A. L. McAlester, and M. L. Rottman. 1984. *The history of the earth's crust.* Englewood Cliffs, NJ: Prentice-Hall.

Head, J. W., C. A. Wood, and T. Mutch. 1976. Geological evolution of the terrestrial planets. *American Scientist* 65:21–29.

Lewis, J. 1974. The chemistry of the solar system. *Scientific American* 230 (March): 60–65.

Pilbeam, D. 1984. The descent of hominoids and hominids. *Scientific American* 250 (March): 84–96.

Siever, R. 1983. The dynamic earth. *Scientific American* 249 (September): 46–65.

Simpson, G. G. 1974. *This view of life.* New York: Harcourt, Brace and World.

Wood, J. A. 1979. *The solar system.* Englewood Cliffs, N.J.: Prentice-Hall.

Worldwatch Institute. 1990. *State of the world 1990.* New York: W. W. Norton.

York, D. 1975. *Planet earth.* New York: McGraw-Hill.

Table 2.1	The Mohs hardness scale
Mineral	**Assigned hardness**
talc	1
gypsum	2
calcite	3
fluorite	4
apatite	5
orthoclase	6
quartz	7
topaz	8
corundum	9
diamond	10

For comparison, the approximate hardnesses of some common objects, measured on the same scale, are: fingernail, 2½; copper penny, 3½; glass, 5 to 6; pocketknife blade, 5 to 6.

A

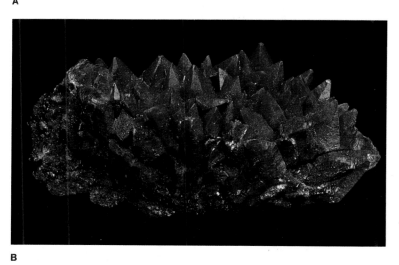

B

Figure 2.7 Here, crystal form mimics the geometry of the internal crystal structure. (*A*) Halite (compare with figure 2.4). (*B*) Calcite (compare with figure 2.5).
© Wm. C. Brown Communications, Inc./Photographs by Bob Coyle

Perhaps surprisingly, **streak,** the color of the powdered mineral, is more consistent from sample to sample than the color of the bulk mineral. Streak is conventionally tested by scraping the sample across a piece of unglazed tile or porcelain and then examining the color of the mark made. There are two principal limitations to the usefulness of streak as an identification tool. First, like color, it is not unique. Many different minerals produce a brown streak, or white, or gray. Second, minerals that are harder than ceramic simply gouge the test surface and are not powdered onto it.

Hardness, the ability to resist scratching, is another physical property that can be of help in mineral identification. Classically, hardness is measured using the Mohs hardness scale (see table 2.1), on which ten minerals are arranged in order of hardness, from talc (the softest, assigned a hardness of 1) to diamond (10). Unknown minerals are assigned a hardness on the basis of which minerals they can scratch and which minerals scratch them. A mineral that scratches gypsum but is scratched by calcite has a hardness of 2½ (the hardness of an average fingernail). Because diamond is the hardest natural substance known, and corundum (aluminum oxide) is the second-hardest mineral, these might be readily identified by their hardnesses.

However, among the thousands of "softer" (more easily scratched) minerals, there are many of any particular hardness, just as there are many of any particular color.

Crystal form, the shape of well-developed crystals of a mineral, is a very useful clue to mineral identification because it is related to the (invisible) internal geometric arrangement or packing of atoms in the crystal structure. In some cases, the relationship is readily apparent (figure 2.7): The ions in sodium chloride are arranged in a cubic structure, and table salt tends to crystallize in cubes; so-called dogtooth crystals of the mineral calcite are similar in shape to the rhombohedral atomic units from which they are built.

However, variations in the conditions under which crystals grow cause different planes of atoms to be more or less prominent in the overall crystal form. A given mineral with a single internal structure can grow in several quite distinctive crystal shapes. Each one necessarily reflects in some way the internal symmetry of the crystal structure, but all crystals of a specific mineral do not necessarily look alike in form. Conversely, different minerals may show the same crystal form (figure 2.8), so crystal form is not uniquely diagnostic of minerals. Moreover, well-developed crystal forms are relatively uncommon in rock and mineral samples, even though all minerals are crystalline internally.

A

B

C

Figure 2.8 Many minerals can share the same crystal form. Here, (A) galena (PbS), (B) fluorite (CaF₂), and (C) halite (NaCl) all form cubes.

Another property controlled by internal crystal structure is **cleavage,** the tendency of minerals to break preferentially in certain directions, corresponding to zones of weakness in the crystal structure. These are related to the nature of the bonding in the mineral. Cleavage can be investigated simply by striking a mineral sample with a hammer. It can be tested even on irregular chunks of mineral that show no well-developed crystals. In some cases, the prominent cleavage directions are the same as the prominent faces of well-formed crystals. Sodium chloride cleaves well in three mutually perpendicular directions that correspond to the faces of its cubic crystals (figure 2.9A). The cleavage fragments of other minerals do not necessarily resemble their well-grown crystals (figure 2.9B). Invariably, however, cleavage directions are directly related to the arrangement of atoms/ions in the crystal structure.

For many minerals, there are no well-defined planes along which the crystals break cleanly. These minerals exhibit more irregular **fracture** rather than cleavage. A distinctive type of fracture is the *conchoidal,* or shell-like, fracture shown by volcanic glass, quartz, and a few other minerals (figure 2.10).

The surface sheen, or **luster,** of minerals is another diagnostic property. Very few terms are used to describe this quality, and they are, for the most part, self-explanatory. Examples include *metallic* (bright and shiny like metal), *vitreous* (glassy), *pearly* (softly iridescent, like pearl), and *earthy* (figure 2.11).

A mineral's **specific gravity** is related to its density. Specific gravity is the ratio of the mass of a given volume of the mineral to the mass of an equal volume of water. By definition, a mineral having the same density as liquid water has a specific gravity of 1. The higher the specific gravity, the denser the mineral. For comparison, sodium chloride has a specific gravity of about 2.16; garnet, 3.1 to 4.2, depending on its exact composition; metallic copper, 8.9; and gold, 19.3. Specific gravity is related to the atomic weights of the

A

Figure 2.9 (*A*) The cleavage directions of halite parallel the cube faces of its crystals. (*B*) Fluorite forms cubic crystals also but cleaves into octahedra.
© Wm. C. Brown Communications, Inc. (*A*) Photograph by Bob Coyle; (*B*) Photograph by Doug Sherman.

B

Figure 2.10 Example of conchoidal fracture and vitreous luster in volcanic glass.
© Wm. C. Brown Communications, Inc./Photograph by Bob Coyle

A

B

Figure 2.11 Examples of (*A*) metallic luster in pyrite and (*B*) earthy luster in clay.
© Wm. C. Brown Communications, Inc. (*A*) Photograph by Doug Sherman; (*B*) Photograph by Bob Coyle.

elements in the mineral; for example, gold is one of the heaviest elements and has a very high specific gravity. The precise determination of specific gravity requires specialized equipment, and many of the most common minerals fall in a narrow range of specific gravity, from about 2.5 to 4. For the nonspecialist, specific gravity is a qualitative tool, most useful for identifying those minerals of unusually high or low specific gravity (density).

Many other properties are useful mainly in identifying those few minerals that possess them. The iron mineral magnetite, as its name suggests, is strongly magnetic, which is very rare among minerals. Sodium chloride, known mineralogically as halite, naturally tastes like table salt; the similar salt potassium chloride (sometimes used in salt substitutes for those on low-sodium diets) also tastes salty but is more bitter. A few minerals *effervesce* (fizz) when acid is dripped onto them. These are carbonate minerals (discussed later in the chapter), which react with the acid and release carbon dioxide gas in the process. Some minerals *fluoresce*, or glow, under ultraviolet light. Fluorite (calcium fluoride) and calcite are examples. Uranium minerals are *radioactive*, which can be detected with a Geiger counter.

Without exact knowledge of a mineral sample's composition and crystal structure, then, one must rely on other less distinctive properties, such as color, hardness, and cleavage. While individual physical properties rarely identify a mineral uniquely, considering a set of properties collectively may narrow the possibilities down to one or a very few minerals. That is, while there may be many green minerals, very few are green *and* show conchoidal fracture *and* glassy luster *and* have a hardness of 6½ to 7 and so on. The more properties that are considered at once, the fewer minerals fit them all.

Types of Minerals

As was indicated earlier, minerals can be grouped or subdivided on the basis of their two fundamental characteristics—composition and crystal structure. In this section, some of the basic mineral groups are introduced briefly. A comprehensive survey of minerals is well beyond the scope of this book, and the interested reader is referred to standard mineralogy texts for more information.

Silicates

The two most common elements in the earth's crust, by far, are silicon and oxygen, which together make up over 70 percent of the crust. It comes as no surprise, therefore, that by far the largest group of minerals is the **silicate** group, all of which are compounds containing silicon and oxygen, and most of which contain other elements as well. Because this group of minerals is so large, it is subdivided on the basis of crystal structure, by the ways in which the silicon and oxygen atoms are linked together.

The basic structural unit of all of the silicates is the *silica tetrahedron,* a compact building block formed by a silicon cation (Si^{4+}) closely surrounded by four

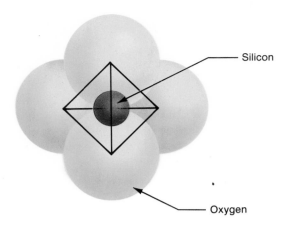

Figure 2.12   The basic silica tetrahedron, building block of all silicate minerals.

oxygen anions (O^{2-}) (figure 2.12). The different silicate mineral groups are distinguished by the way in which these tetrahedra are assembled. A number of specific minerals of various compositions fall into each structural group. Only some of the more common representatives of each structural type are noted here.

The simplest arrangement of tetrahedra is as isolated units. It can be seen arithmetically that an individual SiO_4 tetrahedron has a net negative charge of (-4): $[1(+4) + 4(-2)] = -4$. Thus, in such a silicate, additional cations must balance the electrical charge for the solid to be electrically neutral overall and to bond the negatively charged tetrahedra together. For each tetrahedron, cations totaling four positive charge units are required. In the most common single-tetrahedron silicate, **olivine,** the charges are balanced by iron (Fe^{2+}) and/or magnesium (Mg^{2+}), which requires a total of two $+2$ ions per formula unit: $(Fe,Mg)_2SiO_4$. Another common single-tetrahedron silicate is *garnet,* for which a variety of chemical compositions are possible (recall box 2.2).

Other silicates are known as **chain silicates,** because their silica tetrahedra are arranged in chains formed by the sharing of oxygen atoms between adjacent tetrahedra in one dimension. The most common arrangements are single chains (figure 2.13A) and double chains (figure 2.13B). When tetrahedra share oxygen

atoms, the net effect is less total negative charge from the silica tetrahedra to be balanced by additional cations.

One large group of single-chain silicates is collectively known as **pyroxenes.** This group comprises many specific minerals, differing in the cations involved in the charge-balancing (and in details of crystal structure). A compositionally simple example in which charges are again balanced by iron and magnesium would have the formula $(Fe,Mg)SiO_3$. The common pyroxene *augite* is compositionally more complex, containing calcium and aluminum in addition to iron and magnesium.

The chemical formulas for the double-chain **amphiboles** are even more complex, in part because the amphiboles contain some water in their crystal structures, in the form of hydroxyl ions (OH^-) substituting for oxygen in the tetrahedra. The structural geometry and nature of the charge-balancing are correspondingly more complex. The general formula for a very common amphibole, *hornblende,* illustrates this well: $(Ca,Na)_3(Mg,Fe,Al,Ti)_5(Si,Al)_8O_{22}-(OH,F)_2$.

The chain silicates represent sharing of oxygen between adjacent tetrahedra in one dimension. In the **sheet silicates,** tetrahedra are linked by shared oxygen atoms in two dimensions (figure 2.14). Most sheet silicates contain some water (as hydroxyl ions) also, substituting for oxygen as in the amphiboles.

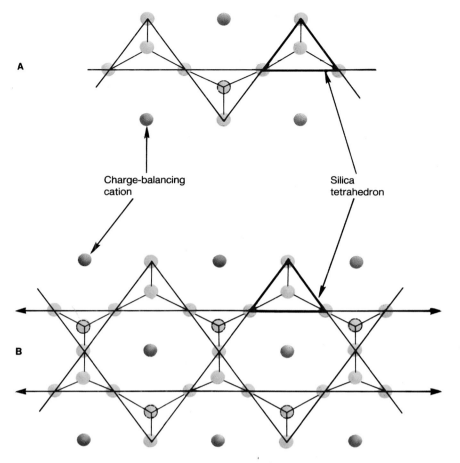

Charge-balancing
cation

Silica
tetrahedron

Figure 2.13 In chain silicates, tetrahedra are linked in one dimension by shared oxygen atoms. (*A*) Single chains, characteristic of pyroxenes. (*B*) Double chains, characteristic of amphiboles.

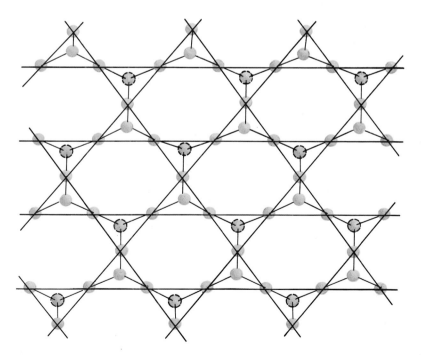

Figure 2.14 The two-dimensional linking of tetrahedra in a sheet silicate.

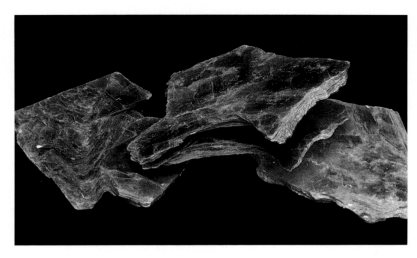

Figure 2.15 Excellent cleavage in micas results from splitting of the crystals between weakly bonded sheets.
© Wm. C. Brown Communications, Inc./Photograph by Bob Coyle

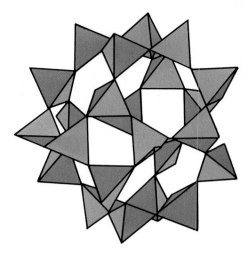

Figure 2.16 When silica tetrahedra are linked in three dimensions, the result is a framework silicate.

The cations commonly fit between stacked sheets, bonding the sheets together. Typically, however, there are too few such cations to hold the sheets very tightly, which is reflected in the macroscopic properties of many sheet silicates. The **micas** are a compositionally diverse group of sheet silicates that have in common excellent cleavage parallel to the weakly bonded sheets of tetrahedra (figure 2.15). Common examples are the light-colored mica *muscovite,* often found in granite, and the dark mica *biotite,* rich in iron and magnesium. *Clay minerals* are also sheet silicates, and their often slippery feeling can likewise be attributed to the sliding-apart of such sheets of atoms. This slippery quality makes some clays useful as lubricants in the drilling muds that cool drill bits in well-drilling rigs. Other clays absorb water and swell, then contract as they dry out, creating unstable slopes and contributing to landslide hazards.

When tetrahedra are firmly linked in all three dimensions by shared oxygen atoms, the result is a **framework silicate** (figure 2.16). If the framework consists entirely of silicon-oxygen tetrahedra, there is no need for further charge-balancing cations in the structure. This is the case with the mineral *quartz,* compositionally the simplest silicate (SiO_2).

There are other framework silicates besides quartz and its polymorphs, however. This is possible because, in silicates of virtually all structural types, some aluminum atoms (Al^{3+}) can substitute in tetrahedra for silicon (Si^{4+}). The resulting charge imbalance must be compensated by additional cations. The most abundant minerals in the crust, the **feldspars,** are such framework silicates in which the charge-balancing cations are sodium (Na^+), potassium (K^+), or calcium (Ca^{2+}).

The feldspars illustrate some of the limitations of solid solution in minerals. Sodium and calcium are very close in size, though slightly different in charge, so they can substitute effectively for each other in a series of sodium-calcium feldspars known as the *plagioclase* feldspars, generalized formula $(Na,Ca)(Al,Si)_2Si_2O_8$. Potassium is so much larger that it does not substitute in plagioclase to any great extent. Conversely, some sodium may substitute in potassium feldspar ($KAlSi_3O_8$) because the charges of sodium and potassium ions are the same ($+1$), although they differ somewhat in size. But calcium differs significantly in both size and charge from potassium, and thus these two elements are not readily interchangeable in the potassium feldspar structure.

Other structural variants occur in the silicates, but less commonly—for example, structures in which tetrahedra are linked into rings or those in which pairs of tetrahedra are joined. In some contexts, it may be useful to refer collectively to a set of silicate minerals of different structures that share a compositional characteristic. For example, the **ferromagnesian** silicates, as the name suggests, comprise those silicates that contain significant amounts of iron (Fe) and/or magnesium (Mg). (They may or may not contain other cations also.) The group includes olivines, pyroxenes, amphiboles, and some micas, among others. The ferromagnesians are commonly characterized by dark colors, most often black, brown, or green. Another possible compositional group is that of the **hydrous** (water-bearing) silicates, which includes micas, clays, amphiboles, and others. A silicate mineral may thus fall within one or more particular compositional groupings in addition to its structural category.

Nonsilicates

Just as the silicates, by definition, all contain silicon plus oxygen as part of their chemical compositions, each nonsilicate mineral group is defined by some chemical constituent or characteristic that all members of the group have in common. Most often, the common component is the same negatively charged ion or group of atoms (somewhat analogous to a silica tetrahedron; a structural unit). Discussion of some of the nonsilicate mineral groups, with examples of more com-

Table 2.2　Some nonsilicate mineral groups*

Compositional class	Compositional characteristic	Examples
carbonates	metal(s) plus carbonate (1 carbon + 3 oxygen atoms, CO_3)	calcite (calcium carbonate, $CaCO_3$) dolomite (calcium-magnesium carbonate, $CaMg(CO_3)_2$)
sulfates	metal(s) plus sulfate (1 sulfur + 4 oxygen atoms, SO_4)	gypsum (calcium sulfate, with water, $CaSO_4 \cdot 2H_2O$) barite (barium sulfate, $BaSO_4$)
sulfides	metal(s) plus sulfur, without oxygen	pyrite (iron sulfide, FeS_2) galena (lead sulfide, PbS) cinnabar (mercury sulfide, HgS)
oxides	metal(s) plus oxygen	magnetite (iron oxide, Fe_3O_4) hematite (iron oxide, Fe_2O_3) corundum (aluminum oxide, Al_2O_3) spinel (magnesium-aluminum oxide, $MgAl_2O_4$)
hydroxides	metal(s) plus hydroxyl (1 oxygen + 1 hydrogen atom, OH)	gibbsite (aluminum hydroxide, $Al(OH)_3$; found in aluminum ore) brucite (magnesium hydroxide, $Mg(OH)_2$; an ore of magnesium)
halides	metal(s) plus halogen element (fluorine, chlorine, bromine, or iodine)	halite (sodium chloride, $NaCl$) fluorite (calcium fluoride, CaF_2)
phosphates	metal(s) plus phosphate group (1 phosphorous + 4 oxygen, PO_4)	apatite ($Ca_5(PO_4)_3F$)
native elements	mineral consists of a single element	gold (Au), silver (Ag), copper (Cu), sulfur (S), graphite (C)

*Other groups exist, and some complex minerals contain components of several groups (carbonate and hydroxyl groups, for example).

mon or familiar members of each, follows. (See also table 2.2.)

The **carbonates** all contain carbon and oxygen combined in the proportions of one atom of carbon to three atoms of oxygen (CO_3). The carbonate minerals all dissolve relatively easily, particularly in acids; the oceans contain a great deal of dissolved carbonate. Geologically, the most important, most abundant carbonate mineral is *calcite,* which is calcium carbonate ($CaCO_3$). Another common carbonate mineral is *dolomite,* which contains both calcium and magnesium in approximately equal proportions ($CaMg(CO_3)_2$).

The **sulfates** all contain sulfur and oxygen in the ratio of 1 to 4 (SO_4). A calcium sulfate—*gypsum*—is the most important, for it is both relatively abundant, and commercially useful for the manufacture of plaster of paris and wallboard.

When sulfur is present without oxygen, the resultant minerals are called **sulfides.** A common and well-known sulfide mineral is the iron sulfide *pyrite* (FeS_2). Pyrite has also been called "fool's gold" because its metallic golden color often deceived early prospectors into thinking that they had

struck it rich. Pyrite is not a commercially useful source of iron because there are richer ores of this metal. Nonetheless, the sulfide group comprises many economically important metallic ore minerals. An example that may be familiar is the lead sulfide mineral *galena* (PbS), which often forms in silver-colored cubes. The rich lead ore deposits near Galena, Illinois, gave the town its name. Sulfides of copper, zinc, and numerous other metals may also form valuable ore deposits.

The **halides** are the minerals composed of metal(s) plus one or more of the halogen elements (the gaseous elements in the next-to-last column of the periodic table: fluorine, chlorine, iodine, and bromine). The most common halide is *halite* ($NaCl$), the most abundant salt dissolved in the oceans.

Minerals that contain just one or more metals combined with oxygen and that lack the other elements necessary for them to be classified as silicates, sulfates, carbonates, and so on are the **oxides.** Iron combines with oxygen in different proportions to form more than one oxide mineral. *Magnetite* (Fe_3O_4) is one of these. Another iron oxide, *hematite* (Fe_2O_3), is sometimes

silvery-black but often has a red color and gives a reddish tint to many rocks and soils. (All colors of hematite have the same characteristic reddish-brown streak, however, illustrating again the usefulness of streak as a diagnostic property in mineral identification.) Many other oxide minerals are known, including *corundum,* the aluminum oxide mineral mentioned earlier.

The **native elements,** as shown in table 2.2, are even simpler chemically than the other nonsilicates. Native elements are minerals that consist of a single chemical element. The minerals are usually named for the corresponding elements. Not all elements can be found, even rarely, as native elements. However, some of the most highly prized materials, such as gold, silver, and platinum, often occur as native elements. Diamond and graphite are both examples of native carbon; here, two mineral names are needed to distinguish these two very different crystalline forms of the same element. Sulfur may occur as a native element, either with or without associated sulfide minerals. Some of the richest copper ores contain native copper. Other metals that may occur as native elements include tin, iron, and antimony.

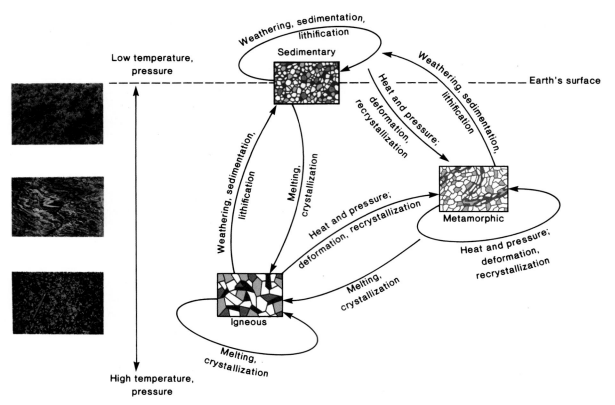

Figure 2.17   The rock cycle (schematic): Geologic processes act continuously to produce new rocks from old rocks.

Rocks and the Rock Cycle

A **rock** is a solid, cohesive aggregate of one or more minerals (or mineral materials, including volcanic glass). This means that a rock consists of many individual mineral grains—not necessarily all of the same mineral—or of mineral grains plus glass, all firmly held together in a solid mass. Because the many mineral grains of a beach sand fall apart when handled, sand is not a rock, although, in time, sand grains may become cemented together to make a rock. Rocks' physical properties are important in determining their suitability for particular applications, such as for construction materials or for the base of a building foundation. Each rock also contains within it a record of at least a part of its history, in the nature of its minerals and in the way the mineral grains fit together. Chapters that follow consider different rock types in detail.

In chapter 1, we noted that the earth is a constantly changing body. Moun-

tains come and go; seas advance and retreat over the faces of continents; surface processes and processes occurring deep in the crust or mantle are constantly altering the planet. One aspect of this continual change is that rocks, too, are always subject to change. We do not have a single sample of rock that has remained unchanged since the earth formed, and many rocks have been changed many times.

Rocks are divided into three classes on the basis of the way in which they form. *Igneous* rocks (chapter 3) crystallize from hot silicate melts. They are common in the deep crust and make up the whole of the mantle. *Sedimentary* rocks (chapter 6) form at the very low temperatures near the earth's surface, as a result of the weathering of preexisting rocks. *Metamorphic* rocks (chapter 7) have been changed (deformed, recrystallized) by the application of moderate heat and/or pressure in the crust. But a rock of one type may be transformed into a rock of another type (or a different rock of the same type) by

suitable geologic processes. A sedimentary rock, for example, may be squeezed, heated, and changed into a metamorphic rock. It may even be heated so much that it melts; when the melt cools and crystallizes, the result is a new igneous rock. If that igneous rock is uplifted on a continent, it can be weathered and redeposited as sediment, to make, in time, new sedimentary rock; and so on.

The idea that rocks are continually subject to change through time is the essence of the **rock cycle** (figure 2.17). The next several chapters treat individual rock types in some detail. Keep in mind, however, that the labels "igneous," "sedimentary," and "metamorphic," necessarily describe a rock only as it has emerged from its last cycle of formation or change. Many earlier stages through which that material has passed have been wholly or partially obliterated by later changes—hence some of the difficulty in piecing together the 4½-billion-year history of the earth from rocks collected today.

Summary

Chemical elements consist of atoms, which are, in turn, composed of protons, neutrons, and electrons. Isotopes are atoms of one element (having, therefore, the same number of protons) with different numbers of neutrons; chemically, isotopes of one element are indistinguishable. Ions are atoms that have gained or lost electrons and thus acquired a positive or negative charge. Atoms of the same or different elements may bond together. The most common kinds of bonding in minerals are ionic (resulting from the attraction between oppositely charged ions) and covalent (involving sharing of electrons between atoms). When atoms of two or more different elements bond together, they form a compound.

A *mineral* is a naturally occurring, inorganic, solid element or compound, with a definite composition (or range in composition) and a regular internal crystal structure. When appropriate instruments for determining composition and crystal structure are unavailable, minerals can be identified from a set of physical properties, including color, crystal form, cleavage or fracture, hardness, luster, specific gravity, and others. Minerals are broadly divided into silicates and nonsilicates. The silicates are subdivided into structural types (for example, chain silicates, sheet silicates, framework silicates) on the basis of how the silica tetrahedra are linked in each mineral. Silicates may alternatively be grouped by compositional characteristics. The nonsilicates are subdivided into several groups, each of which has some compositional characteristic in common; examples include the carbonates (each containing the CO_3 group), the sulfates (SO_4), and the sulfides (S).

Rocks are cohesive mineral aggregates. All rocks are part of the rock cycle, through which old rocks are continually being transformed into new ones. A consequence of this is that no rocks have been preserved throughout earth's history, and many early stages in the development of any one rock may have been erased by subsequent events.

Terms to Remember

amphibole	ion
anion	ionic bond
atom	isotope
atomic mass number	luster
	mica
atomic number	mineral
carbonates	mineraloid
cation	native element
chain silicates	neutron
cleavage	nucleus
compound	olivine
covalent bond	oxide
crystal form	polymorph
crystalline	proton
electron	pyroxene
element	radioactivity
feldspars	rock
ferromagnesian	rock cycle
fracture	sheet silicate
framework silicates	silicates
	solid solution
glass	specific gravity
halides	streak
hardness	sulfate
hydrous	sulfide
inert	

Questions for Review

1. What are *isotopes?*
2. Compare and contrast ionic and covalent bonding.
3. How is a *mineral* defined? What are the two key identifying characteristics of a mineral?
4. What is the phenomenon of solid solution, and how does it affect the definition of a mineral?
5. Explain the limitations of the use of color as a tool in mineral identification. Cite and explain any three other physical characteristics that might aid in mineral identification.
6. What is the basic structural unit of all silicate minerals? Describe the basic structural arrangements of chain silicates, sheet silicates, and framework silicates.
7. Give the compositional characteristic common to each of these nonsilicate mineral groups: carbonates, sulfides, oxides.
8. What is a *rock?*
9. Describe the basic concept of the rock cycle.

For Further Thought

1. More ancient rocks generally are less widely found and more difficult to interpret than are younger rocks. Explain why, in the context of the rock cycle.
2. Choose one of the following minerals or mineral groups, and investigate its range of physical properties and its applications: quartz, calcite, garnet, clay.

Suggestions for Further Reading

Berry, L. G., B. Mason, and R. V. Dietrich. 1983. *Mineralogy.* 2d ed. San Francisco: W.H. Freeman.

Dana, J. D. 1977. *Manual of mineralogy.* 19th ed. Revised by C. S. Hurlbut, Jr. and C. Klein. New York: Wiley.

Deer, W. A., R. A. Howie, and J. Zussman. 1978. *Rock-forming minerals.* 2d ed. New York: Halstead.

Dietrich, R. V., and B. J. Skinner. 1979. *Rocks and rock minerals.* New York: John Wiley and Sons.

Ernst, W. G. 1969. *Earth materials.* Englewood Cliffs, N. J.: Prentice-Hall.

Hurlbut, C. S., Jr. 1968. *Minerals and man.* New York: Random House.

Kirklady, J. F. 1972. *Minerals and rocks in color.* New York: Hippocrene Books.

O'Donoghue, M. 1976. *VNR color dictionary of minerals and gemstones.* New York: Van Nostrand Reinhold.

Philips, W. J., and N. Philips. 1980. *An introduction to mineralogy for geologists.* New York: John Wiley and Sons.

Sinkankas, J. 1964. *Mineralogy for amateurs.* Princeton, N. J.: D. Van Nostrand.

Watson, J. 1979. *Rocks and minerals.* 2d ed. Boston, Mass.: Allen and Unwin.

Zoltai, T., and J. H. Stout. 1984. *Mineralogy: Concepts and principles.* Minneapolis, Minn.: Burgess.

CHAPTER 3

Outline

Igneous Rocks and Processes

The vast expanses of granite exposed in
Yosemite National Park were crystallized deep
in the crust, then raised by tectonic processes
and exposed by the slow wearing-away of
overlying rocks.

Introduction

The term **igneous** comes from the Latin *ignis,* meaning "fire." The name is given to rocks formed at very high temperatures, crystallized from a molten silicate material known as **magma.** This chapter examines magmas and aspects of their formation and crystallization; basic characteristics and classification of igneous rocks; and some of the structures magmas form in the crust. The next chapter focuses on the surface manifestations of igneous activity—volcanoes and related phenomena.

Figure 3.1   This magma contained gas bubbles when it solidified.
© Wm. C. Brown Communications, Inc./Photograph by Bob Coyle

Origin of Magmas

High temperatures are required to melt rock, and within the earth, temperatures increase with depth. The rate of temperature increase, called the **geothermal gradient,** averages about 30° C per kilometer (85° F per mile) of depth in the continental crust. Magmas originate at depths where temperatures are high enough to melt rock, usually in the upper mantle, at depths between about 50 and 250 kilometers. However, the required temperatures (and depths) vary considerably, depending on several other factors.

Effects of Pressure

As the temperature of a solid is increased, its individual atoms vibrate ever more vigorously, until their energy is sufficient to break the bonds holding them in place in the solid structure. They then flow freely, in a disordered liquid.

Most substances are more dense in their crystalline solid form than in the liquid state, and thus an increase of pressure favors the more compact, solid arrangement of atoms. At high pressure, then, correspondingly higher temperatures are required to impart enough energy to the atoms to cause melting. This explains why most of the earth's interior is not molten. Temperatures are indeed high enough through

most of the earth to melt the rocks, if they were at atmospheric (surface) pressures. But the weight of overlying rock puts enough pressure on the rocks below that most remain solid even at temperatures of thousands of degrees. If that pressure is lessened, for example by opening of fractures in rocks above, the solid may begin to melt. Parts of the uppermost mantle are molten; the deeper mantle, though hotter, is solid because of the effect of pressure on melting temperature.

The effects of pressure on melting can also be illustrated by a commonplace example involving ice. Ice is unusual in that, at least close to its melting point, it is *less* dense than water (note that ice cubes and icebergs float). An increase in pressure favors, as usual, the denser form—in this case, water. This is what makes ice skating possible: Concentrating a body's weight on a narrow skate blade puts tremendous pressure on the ice below, and a little of it melts, providing a watery layer on which the skate can glide smoothly.

Effects of Volatiles

Natural magmas contain dissolved volatiles—water and various gases (oxygen, carbon dioxide, hydrogen sulfide, and others). Sometimes, this is

obvious from the resultant rock, which may preserve bubbles formed by gases in the magma (figure 3.1). The general effect of dissolved volatiles is to lower the melting temperatures of silicate minerals. This is illustrated in figure 3.2. Increasing the water pressure drives more water into the melt and lowers the melting temperature of sodic plagioclase, most dramatically for small additions of water; the effect decreases for high water pressure. Exactly how much melting temperatures are lowered in the presence of volatiles varies with the nature of the volatiles and the minerals being melted, but the general principle holds in any case: more volatiles, lower melting temperatures.

Effects of Other Solids Present

When two or more different minerals are in contact, the presence of each lowers the melting temperature of the other, to a point. This is why salt can be used to melt ice on a sidewalk in winter. The presence of the salt lowers the melting temperature of the ice so that the ice melts below the freezing point of pure water (0° C, or 32° F). (Presumably, the ice lowers the melting temperature of the salt, too, but salt has such an extremely high melting point

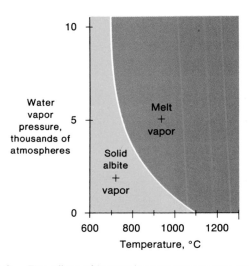

Figure 3.2 The effect of increasing water pressure on the melting temperature of sodic plagioclase (albite).

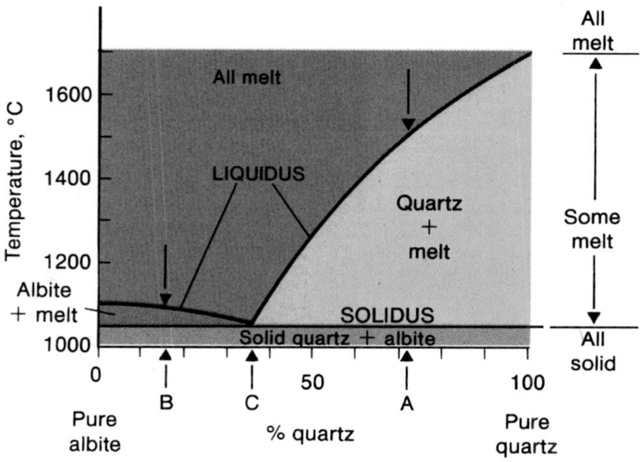

Figure 3.3 A mix of quartz and albite can melt at a lower temperature than either mineral alone.

to begin with that it remains solid unless, of course, it simply dissolves in the melted ice.)

Similar effects are observed with minerals. Figure 3.3 is a melting diagram for mixtures of quartz and sodic plagioclase (albite). While the quartz in this case appears to have little effect on the crystallization of albite, the effect of adding albite to the quartz is dramatic, reducing the melting temperature of quartz from 1713° C (3115° F) to about 1100° C (2010° F) for a mix of 40 percent quartz, 60 percent albite.

The curved lines in figure 3.3 are *liquidus* curves, which separate conditions under which the system is entirely molten from conditions under which some solid is present. Crystallization of a given melt, as it cools, begins at the liquidus temperature for that composition. (For example, for composition A, which is 70 percent quartz, the liquidus temperature is 1470° C.) The horizontal line at about 1060° C is the *solidus* and represents the temperature below which the system is entirely solid. Melting relationships are the reverse of the crystallization process: The melting of a warmed rock consisting of quartz plus albite begins at the solidus temperature, 1060° C, and continues up to the liquidus temperature for that rock's composition.

Since the vast majority of rocks contain many minerals, it is safe to expect melting of the rocks at temperatures somewhat below those at which the pure minerals would melt, but the relationships are likely to be complex, and it may not be possible to predict each rock's exact melting temperature. Note also that melting in natural rocks typically occurs over a range of temperatures, with different minerals melting at different temperatures. The transition from a wholly solid to completely molten material may span several hundred degrees, and, in fact, complete melting may not be necessary for the magma to mobilize and flow. Many magmas are a sort of mush of crystals suspended in silicate liquid.

Crystallization of Magmas

Once an appropriate combination of factors has produced a quantity of melt, the melt tends to flow away from where it was produced. Usually, it flows upward, into rocks at lower temperatures and pressures. As it moves upward, it cools. The cooling, in turn, leads to crystallization, as the atoms slow down and eventually settle into the orderly arrays of crystals. The details of the rock thus produced vary with the composition of the melt and the conditions under which it crystallized. However, some generalizations about the resulting mineralogy and texture are possible.

Sequence of Crystallization

As already observed, a mix of minerals melts over a range of temperatures. Likewise, a magma crystallizes over a range of temperatures, or in other words, over some period of time during cooling, as different minerals begin to crystallize at different temperatures/times. Moreover, because most magmas originate in the upper mantle, they are in a very general sense similar in composition initially, consisting predominantly of silica (SiO_2), with varying lesser proportions of aluminum, iron, magnesium, calcium, sodium, potassium, and additional minor elements. Magmas therefore tend to follow a predictable crystallization sequence, in terms of the principal minerals forming. However, the proportions of these minerals in the ultimate rock vary, and igneous rocks, overall, span a broad range of compositions.

The basic crystallization sequence was recognized more than half a century ago by geologist Norman L. Bowen, who combined careful laboratory experiments on compositionally simple silicate systems with wide-ranging field observations of the more complex natural rocks. The result, known as **Bowen's reaction series,** is illustrated in figure 3.4. Those minerals that tend to crystallize at higher temperatures are shown nearer the top of the series; the later-crystallizing, low-temperature minerals are near the

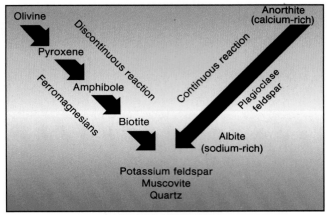

Figure 3.4 Bowen's reaction series.

bottom. In general, the earliest minerals to crystallize are relatively low in silica, so that the residual magma remaining after their crystallization is more enriched in silica relative to its starting composition. The high-temperature portion of the sequence is also subdivided into a ferromagnesian branch and a branch involving plagioclase feldspar.

The plagioclase branch is a **continuous reaction series.** This refers to interaction between crystals already formed and the remaining melt. Recall from chapter 2 that plagioclase is a solid solution between a calcium-rich endmember (anorthite, $CaAl_2Si_2O_8$) and a sodium-rich endmember (albite, $NaAlSi_3O_8$). The more calcic compositions are the higher-temperature end of the series: Pure anorthite melts at about 1,550° C, pure albite at 1,100° C. Sodium and calcium are freely interchangeable in the plagioclase crystal structure. The first plagioclase to crystallize from a magma, at high temperatures, is a rather calcic one; but as the magma cools, the crystals react continuously with the melt, with more and more sodium entering into the plagioclase, but no changes in basic crystal structure. (Note that sodic plagioclase also contains a higher proportion of silicon, so the later plagioclases are more silica-rich, too.) If cooling is too rapid for complete reaction between crystals and melt during cooling, the resultant crystals show

concentric compositional zones, with calcium-rich cores grading outward through progressively more sodium-rich compositions.

The ferromagnesian side of the crystallization sequence is a **discontinuous reaction series.** Olivine is the first of the ferromagnesians to crystallize. After a period of crystallization, the olivine is so out of balance chemically with the increasingly silica-rich residual melt that olivine and melt react to form pyroxene. (Recall that the ratio of [iron plus magnesium] to silicon in olivine is 2 to 1, while in pyroxene it is about 1 to 1.) Assuming that there is sufficient silica available, all of the olivine is converted to pyroxene at that point. After an interval of pyroxene crystallization, the pyroxene, again out of chemical balance with the remaining melt, reacts with it, and pyroxenes are converted to amphiboles, and so on. The last of the common ferromagnesians to crystallize is biotite mica. There is a progression in structural complexity, too, from the simplest silicate structure (olivine) through chains to a sheet structure (biotite). The discontinuous reaction series, then, is marked by several changes in mineralogy/crystal structure during cooling and crystallization.

At the end of the crystallization sequence, at lowest temperatures, any last, silica-rich residual melt crystallizes potassium feldspar, muscovite mica, and quartz. Note that the hydrous sil-

icates—amphiboles and micas—are relatively late in the sequence. At very high temperatures, hydrous minerals are unstable, and any water stays in the melt. Also, not every magma progresses through the whole sequence. A more **mafic** magma (rich in magnesium and iron, poorer in silicon) will be completely crystallized before the lattermost stages of the sequence are reached; it does not start with sufficient silica to make quartz at the end. A very **silicic** (silica-rich, iron- and magnesium-poor) magma will indeed reach the final stages, by which time reactions with the melt have eliminated early-formed olivine, pyroxene, and calcic plagioclase. Mafic magmas, then, produce rocks rich in the minerals near the top of the diagram in figure 3.4; silicic magmas produce rocks that are dominated by the minerals near the bottom and that are poor in ferromagnesians. The latter rocks are typically rich in feldspar and quartz (silica) and are therefore also called **felsic.**

Modifying Melt Composition

The previous discussion tacitly assumed that each crystallizing magma behaves as a *closed system,* neither gaining nor losing matter. This is often not the case in natural systems. Magma compositions may be modified after the melt is formed. The result is a product somewhat different from what would be expected on the basis of the original melt composition.

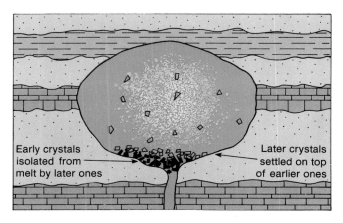

Figure 3.5 Crystal settling in a magma chamber as a mechanism of fractional crystallization. Early crystals are isolated from the residual melt by later ones.

Early crystals isolated from melt by later ones

Later crystals settled on top of earlier ones

Fractional crystallization is the primary way in which melt composition can be changed. In this process, early-formed crystals are physically removed from the remaining magma and so are prevented from reacting with it. One way in which this can happen is if the early crystals settle out of the crystallizing magma and are isolated from the melt by later crystals settling above them (figure 3.5). The remaining melt may even move away through zones of weakness in surrounding rocks, leaving the early crystals behind altogether. The result is that the average composition of the melt, minus its early, low-silica minerals, is shifted toward a more silica-rich, iron- and magnesium-poor composition. That melt can then progress further down the crystallization sequence than would have been possible if all crystals had remained in the melt as it cooled. A great diversity of rock compositions can be produced from a single starting magma composition by varying the extent and timing of fractional crystallization during cooling.

Overall magma composition also can be changed if a magma **assimilates** the rock around it, incorporating pieces of it, melting and mixing it in (figure 3.6). Given the relative temperatures at which mafic and felsic magmas crystallize, assimilation occurs most readily when the initial melt is a (hotter) mafic magma and the assimilated material is more silicic. The result is a modified magma of somewhat more silicic com-

Figure 3.6 Assimilation was in progress here when the melt solidified.

position, which could move correspondingly further down the crystallization sequence.

Magma mixing, in which two melts combine to produce a hybrid melt intermediate in composition between them, is another way in which melt composition can be changed. Because unusual geologic conditions are needed to produce two distinctly different magmas in nearly the same place at the same time, however, magma mixing is rather rare.

Classification of Igneous Rocks

The most fundamental division of igneous rocks is made on the basis of depth of crystallization, as reflected in texture.

Textures of Igneous Rocks

The most noticeable textural feature of most igneous rocks is the *grain size,* the size of the individual mineral crystals.

Figure 3.7 A glassy igneous rock. This melt cooled so rapidly that there was no time for crystals to form.

An important control on grain size, as indicated in chapter 2, is cooling rate. If a magma cools slowly, there is more time for atoms to move through the melt and attach themselves at suitable points on growing crystals. The resulting rock is coarser-grained. A **phaneritic** igneous rock is one in which individual crystals are large enough to be readily visible to the naked eye. In a rapidly cooled magma, there is less time for crystal growth, so the rock is finer-grained. The individual crystals may not be seen easily with the unaided eye; the resulting texture is called **aphanitic.** In extreme cases of rapid cooling and limited crystal growth, the result is a *glassy* rock, with no obvious crystals (figure 3.7). Some quickly cooled magmas also trap bubbles or pockets of gas, which are termed **vesicles;** the corresponding texture is described as *vesicular* (recall figure 3.1).

Some igneous rocks have a two-stage cooling history, with an initial stage of slow cooling that allows some large crystals to form followed by a stage of rapid cooling that leaves the rest of the rock (**groundmass**) finer-grained. This might happen, for example, if a magma began to crystallize slowly at depth and then was suddenly erupted from a volcano. The resulting rock is called a **porphyry;** the texture is described as *porphyritic* (figure 3.8; see also box 3.1). The coarse crystals embedded in the finer groundmass are termed **phenocrysts,** the prefix *pheno-* coming from the Greek for "to show"—in other words, these are very obvious crystals.

A

B

Figure 3.8 (*A*) The phenocrysts in this porphyry formed early, during a period of slow cooling. Later, rapid cooling caused the remaining melt to crystallize in a mass of smaller crystals. (*B*) Though it is less common, coarse-grained rocks can be porphyritic also; here, coarse feldspar phenocrysts in granite.

(*A*) © Wm. C. Brown Communications, Inc./Photograph by Bob Coyle

BOX 3.1

Looking at Rocks Another Way: Thin Sections

Volcanic rocks are so fine-grained that it is commonly impossible to distinguish individual grains with the naked eye. With the majority of rocks of all types, even if individual mineral grains can be seen, it is hard to examine in detail the grain shapes, internal characteristics of individual crystals, and so forth. Geologists therefore rely on microscopic examination of **thin sections,** paper-thin slices of rock mounted on glass slides, to learn more than they can from looking at a chunk of rock unaided.

A thin section is typically ground down to a thickness of about 0.03 millimeters (0.0012 inches), at which thickness most minerals are transparent or translucent (figures 1A and 1B). The special microscopes for examining thin sections use polarized light (light oriented so that all rays vibrate in parallel). The light rays are deflected or rotated in passing through crystals, and different minerals produce different deflections. If the thin section is sandwiched between two polarizers at

right angles to each other, *interference colors* diagnostic of the different minerals are seen; compare figure 1C. A suitably trained geologist would be able to recognize plagioclase (gray-and-white striped) and pyroxene (colored) phenocrysts in this volcanic rock, and smaller crystals of both, plus some glass and magnetite (the latter opaque even in plain light) in the groundmass.

A

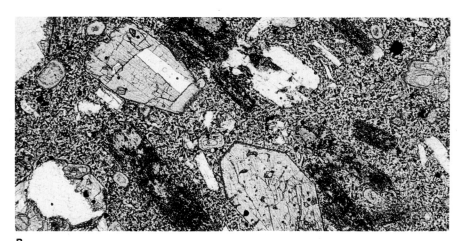

B

C

Figure 1 (*A*) Thin section of porphyritic andesite. Note how transparent it is. (*B*) Microscopic view of the thin section, in polarized light. (*C*) The same thin section viewed between crossed polarizers.

Figure 3.9 Pegmatite, a very coarse-grained igneous rock.

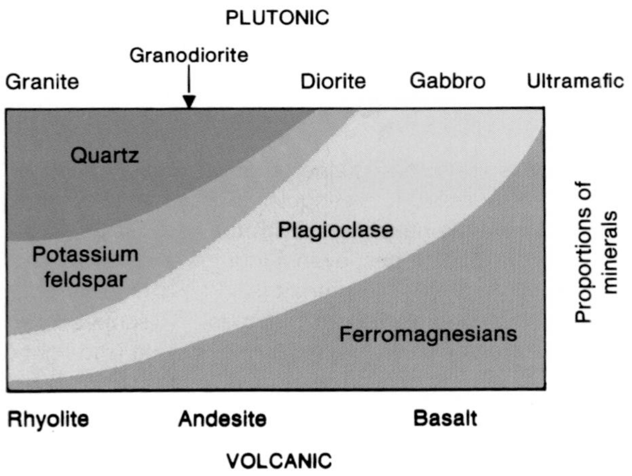

PLUTONIC

Granodiorite

Granite   Diorite   Gabbro   Ultramafic

Quartz

Potassium
feldspar

Plagioclase

Ferromagnesians

Proportions of
minerals

Rhyolite   Andesite   Basalt

VOLCANIC

Figure 3.10 A simplified classification of igneous rocks, based on mineralogy (related to chemical composition) and origin (plutonic versus volcanic).

Grain size also can be affected by melt composition. Silicic melts are more viscous, or thicker, than mafic ones. In all silicate melts, silica tetrahedra exist in the melt even before actual crystallization. In a mafic melt, most of these tetrahedra float independently. In more silica-rich melts, the tetrahedra are more extensively linked. As a result, atoms move less freely through the melt, and more time is required for atoms to move into position on growing crystals. Most volcanic glasses are silicic in composition for this reason (even when dark-colored). Such melts are so stiff and viscous that rapid cooling produces, not small crystals, but virtually no crystals at all, as in figure 3.7.

On the other hand, some late-stage, residual melts, left over after extensive fractional crystallization, have accumulated high concentrations of dissolved volatiles, so they are quite fluid, and atoms move easily through them. Even with moderate cooling rates, very large crystals can grow in a volatile-rich, fluid magma. The resultant extremely coarse-grained rock is called a **pegmatite** (figure 3.9). The coarse crystals and often unusual compositions of some pegmatites make them important sources of gemstones. Examples of gems mined from pegmatites include emerald, aquamarine, tourmaline, and topaz. Individual crystals in some pegmatites can be several meters long.

The igneous rocks are broadly subdivided into two categories on the basis of depth of crystallization. **Plutonic** igneous rocks are those crystallized at some depth below the surface; they take their name from Pluto, the Greek god of the lower world. Their exposure at the surface implies the erosion of the rocks that overlay them at the time of crystallization. Rocks are poor heat conductors, so magmas at depth are well insulated and cool slowly. Plutonic rocks, then, are generally recognized on the basis of their coarse grain sizes. **Volcanic** rocks are those formed from magmas cooled at or near the surface. They are typically fine-grained, aphanitic or glassy. Porphyritic rocks with very fine-grained groundmass are also considered volcanic. Further subdivisions within the textural classes are made primarily on the basis of composition, which permits a specific rock name to be assigned. A summary of principal igneous rock types and their corresponding compositions is shown in figure 3.10 and explained in the sections that follow.

Plutonic Rocks

The coarse grain sizes of plutonic rocks make their preliminary identification relatively simple even without special equipment. They are classified on the basis of relative proportions of certain light and dark (usually ferromagnesian) minerals. A rock consisting entirely of ferromagnesians and dark, calcic plagioclase feldspar is a **gabbro** (figure 3.11). (In the extreme case where feldspar is virtually absent and the rock consists almost wholly of olivine and pyroxene, it is termed **ultramafic.**) A rock that is somewhat more silica-rich than gabbro contains some lighter-colored sodic plagioclase and potassium feldspar, with the mix of light and dark minerals giving it a salt-and-pepper appearance. This is **diorite** (figure 3.12). If the rock is sufficiently silica-rich that appreciable quartz is present, and the proportion of ferromagnesians is correspondingly less, the rock is a **granite** (figure 3.13; see also figure 3.8B).

Although plutonic rocks are put in distinct categories, they actually span a continuum of chemical and mineralogical compositions (see figure 3.10). The boundaries between categories are somewhat arbitrary. One can indicate that a rock is of an intermediate com-

Figure 3.11 Gabbro, a mafic plutonic rock; here, a plagioclase-rich example.
© Wm. C. Brown Communications, Inc./Photograph by Bob Coyle

Figure 3.12 Diorite, an intermediate plutonic rock with salt-and-pepper coloring.
© Wm. C. Brown Communications, Inc./Photograph by Doug Sherman

Figure 3.13 Granite, a silica-rich plutonic rock.

position by using a hybrid name for it: For example, a dioritic rock containing just a little quartz could be called a *granodiorite*, to indicate that its composition lies between the quartz-rich granite and quartz-free diorite classes.

Many rock types were named before the development of sophisticated chemical analytical methods and before there was general agreement on nomenclature. This has left some confusing contradictions among the rock names. The name *gabbro* comes from the locality Gabbro, Italy, but by modern classification, the plutonic rocks found there would now be called diorite, not gabbro!

Volcanic Rocks

Most plutonic rock types have fine-grained volcanic compositional equivalents, as can be seen in figure 3.10. Volcanic rocks are more difficult to identify definitively in handsample, for the individual mineral grains are, by definition, tiny. Color is the most frequently used means of identification in the absence of magnifying lenses or other equipment, but as with mineral identification, it is not an entirely reliable guide. Examination of thin sections (see box 3.1) is often critical to the proper identification of volcanic rocks.

The most common volcanic rock is **basalt** (figure 3.14), the fine-grained equivalent of gabbro. This dark-colored (usually black) rock makes up the sea floor and is erupted by many volcanoes on the continents as well. **Andesite** is the name given to volcanic rocks of intermediate composition. They are typically lighter in color than basalt, often green or gray (see figure 3.8A). The volcanic equivalent of granite is **rhyolite** (figure 3.15). Some rhyolites are pinkish in color, but many are not. Distinguishing among the light-colored andesites and rhyolites can be difficult unless the rhyolite is porphyritic and quartz is visible among the phenocrysts.

Figure 3.14 Basalt, the volcanic equivalent of gabbro and the most common volcanic rock type.

Figure 3.15 Rhyolite, a silicic volcanic rock chemically equivalent to granite. Sample at left is porphyritic.

© Wm. C. Brown Communications, Inc./Photograph by Bob Coyle

A few volcanic rock names lack compositional implications. An example is **obsidian,** volcanic glass, which was shown in figures 3.7 and 2.10. The term obsidian can be applied to glassy rock of any composition, although in fact, most obsidians are rhyolitic in composition.

As with plutonic rocks, mixed names can denote rocks with compositions overlapping two compositional classes (for example, basaltic andesite). A textural and a compositional term can also be combined to provide a more complete description of the rock (for example, porphyritic andesite, vesicular basalt).

Further Considerations

Specialists in igneous rocks naturally use a far more elaborate nomenclature than discussed here, with many more specialized terms. Many unusual rock types could not properly be classified in any of the simple categories previously discussed. The foregoing scheme, however, covers the vast majority of igneous rocks and provides enough of a working vocabulary for discussion of the geologic processes to be dealt with later in the text.

Intrusive Rock Structures

Magma moving through the crust is intruding the surrounding rock, which is commonly termed the **country rock,** or **wallrock.** If the magma erupts at the earth's surface, it has become *extrusive.* Extrusive rocks and structures (for example, volcanoes) are discussed in some detail in chapter 4. Here, we briefly review the more common forms and features of *intrusions,* igneous rock masses formed by magma crystallizing below the earth's surface.

Factors Controlling the Geometry of Intrusions

Both the properties of the magma and the properties of the surrounding rocks play a role in determining the shape of the intrusion formed. Magmas vary in density, with the iron-rich mafic magmas the more dense. Denser magmas are less buoyant with respect to the country rock, and their extra mass may even cause the country rock to sag around the magma body. The viscosity of the magma influences how readily it flows through narrow cracks or other openings in the country rock. This can be demonstrated in the kitchen using a sieve: Water pours rapidly through the sieve; honey or molasses, which are more viscous, seep through more slowly; and honey that has partially crystallized (analogous to the crystal-liquid mush of a partly solidified magma) may not pass through the sieve at all, instead remaining as a coherent mass within it. Similarly, very fluid magmas can move through quite narrow cracks in rocks, while thick, viscous magmas are more likely to remain in a compact mass. Sometimes, too, the magma is under unusual pressure from trapped gases within it, which allows it to intrude forcibly wallrocks that it might ordinarily be unable to penetrate.

The strength of the wallrock, and whether or not it is fractured, also influence the shape of intrusions. Fractures in rock are zones of weakness through which magmas can pass more easily. Zones of weakness may also exist at a contact between dissimilar rock types in the wallrock. The shape of an intrusive body may be controlled by the geometry of zones of weakness in the wallrock through which the magma flows preferentially.

Forms of Intrusive Bodies

Pluton is a general term for any body of plutonic rock, that is, any igneous rock mass crystallized below the earth's surface. The term has no particular geometric significance. Plutons are classified according to their shapes and their relationship to any structures in the country rock. A pluton must necessarily postdate (be younger than) the country rock that it intrudes; this is a basic principle of relative dating of rocks, discussed further in chapter 8. The pluton is said to be **concordant** if its contacts are approximately parallel to any structure (such as compositional layering or folds) in the country rock; it is said to be **discordant** if its contacts cut across the structure of the country rock. A few examples should clarify the distinction.

Cylindrical plutons, elongated in one dimension (usually the vertical), are **pipes** or **necks** (figure 3.16). These are believed to be the feeders or conduits that once carried magma up to a volcano above them; erosion has since removed the volcano and much of the country rock around the pipe. Pipes are typically discordant.

Tabular, relatively two-dimensional plutons commonly result from magma emplacement along planar cracks or

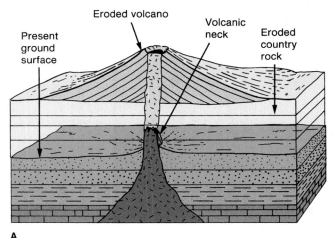

A

B

Figure 3.16 A volcanic neck or pipe—all that remains of a long-eroded volcano. (*A*) Schematic diagram: Note cylindrical shape and discordant character. (*B*) An example exposed at the surface by erosion: Devil's Tower, Wyoming.

zones of weakness. They are called **dikes** if they are discordant, **sills** if they are concordant (figure 3.17).

Concordant plutons that are more nearly equidimensional are less common (figure 3.18). Those that have

flat floors, or bottoms, and that have caused doming or arching of the rocks above them are termed **laccoliths.** Those that have floors that are concave upward (their tops may not be visible) are called **lopoliths.** Laccoliths are

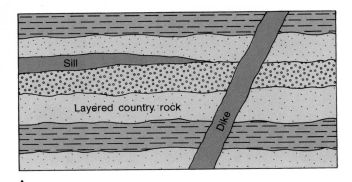

A

B

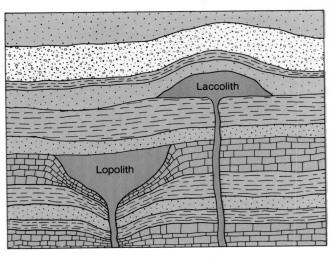

Figure 3.18 Concordant three-dimensional plutons: a laccolith and a lopolith.

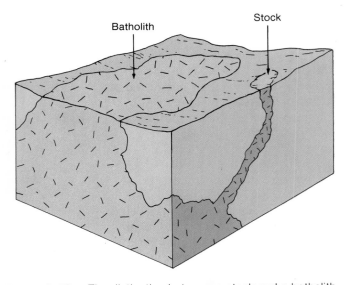

Figure 3.19 The distinction between a stock and a batholith is purely one of scale of exposure at the surface. Both are discordant three-dimensional plutons.

Figure 3.17 Tabular plutons. (*A*) Sketches of a concordant sill and a discordant dike. (*B*) Example of a dike.

commonly formed by silicic rocks (magmas), lopoliths by mafic ones. This suggests that the density of the magma has played a significant role in shaping the pluton, the lopolith perhaps showing sagging of the country rock under the weight of dense mafic rocks.

Discordant equidimensional plutons are somewhat arbitrarily divided on the basis of the area of rock exposed at the surface. A **stock** is exposed over less than 100 square kilometers (about 35 square miles), while a **batholith** is anything larger.

Actually, many stocks may just be small bulges of magma fingering upward from a batholith below, not yet exposed at the surface by erosion (figure 3.19), so the distinction is not particularly important. Very large batholiths, which can cover thousands of square kilometers of outcrop area and extend to a depth of 5 kilometers (3 miles) or more, are multiple intrusions. Many batches of magma were emplaced to form them, and often, many smaller plutons can be distinguished within the large batholith.

Other Features of Intrusions

A pluton may exhibit various textural features, regardless of its overall geometry.

When a hot magma intrudes much cooler country rock, the melt near the contacts is quenched, or rapidly cooled. The resultant **chilled margins** are recognizable because they are finer-grained than the interior of the pluton (figure 3.20). Chilled margins are more commonly found in shallower plutons (and volcanic rocks) because the tem-

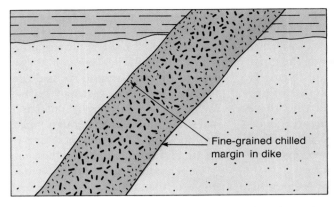

Figure 3.20 Chilled margin. Rapid cooling at the edges of a pluton results in a fine-grained chilled margin.

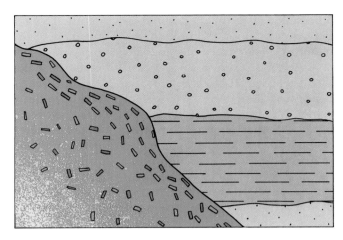

Figure 3.21 Flow texture may be shown by parallel alignment of elongated crystals in an igneous rock.

perature contrast between magma and country rock is greater at shallower depths in the crust.

Plutons also sometimes show compositional layering. It usually develops as a result of gravitational settling-out of denser crystals to the floor of the magma chamber as crystallization proceeds; recall figure 3.5. Layering is more frequently observed in mafic plutons, probably for two reasons. First, in the early stages of crystallization, mafic magmas crystallize abundant dense ferromagnesian minerals, which are especially prone to settling. Second, settling occurs more readily in a less viscous liquid, and mafic magmas are generally less viscous. (To illustrate the effect of viscosity, drop a pebble or dried pea into a glass of water and a glass of molasses, and compare the results.)

If some crystals have formed before the magma stops flowing actively, evidence of that flow may be found in the parallel alignment of platy or elongated crystals (figure 3.21). Near the edges of a pluton at least, the flow direction indicated is generally parallel to the contacts. Bubbles, like crystals, may also be aligned during flow.

As magma invades, bits of country rock may be caught up in the molten mass. Some may be completely assimilated, but some may be preserved as **xenoliths,** as in figure 3.6. Xenoliths take their name from the Greek *xenos* ("stranger") and *lithos* ("stone"); they are bits of alien rock engulfed in a genetically unrelated magma.

Figure 3.22 The vast expanse of granite in Yosemite National Park is typical of a large batholith.

Batholiths and the Origin of Granite

Most batholiths are granitic to granodioritic in composition. What is the source of such large volumes of granitic magma (figure 3.22)?

As mentioned earlier, most magma originates in the upper mantle, where temperatures are high enough and pressures simultaneously low enough to permit melting. What occurs is actually **partial melting,** in which those minerals with the lowest melting points do melt, while higher-temperature minerals do not. Inspection of Bowen's reaction series (figure 3.4) indicates that the first minerals to melt with increasing temperature are the major constituents of typical granite. However, the mantle is ultramafic in overall composition; even the very earliest melt would not be granitic, and by the time a significant fraction of the mantle had melted, the melt would be basaltic in composition through the addition of melted ferromagnesians and plagioclase. Even if a very small percentage of melt were granitic in composition, there would be a twofold problem: first, that a huge volume of mantle would then have to be involved to account for the immense volume of granite in a major batholith, and second, that it is difficult, mechanically, to squeeze a percent or two of melt out of almost completely solid mantle, in order to emplace it into the overlying crust.

An alternative way to make granitic magma would be to start with the basaltic magma expected with significant melting of an ultramafic mantle, then subject it to fractional crystallization at depth. This could proceed, removing the more mafic constituents, until the residual melt was granitic, at which point the melt could intrude the crust. Again, there is a volume problem. To end up with the large volumes of granitic melt represented by batholiths, it would be necessary to start with enormous volumes of basaltic melt. There is little evidence of such extensive melting in the upper mantle.

A third possibility for generating granitic magma is to involve some crust in the melt. For instance, the continental crust is, on average, granodioritic in composition. Partial or complete melting of continental crust could produce magma of the required composition. So would assimilation of considerable continental crust by a rising basaltic, mantle-derived magma. In either case, the problem is one of heat budget, for the continental crust is not normally hot enough to melt, even at depth. Granitic rocks do melt at lower temperatures than do mafic rocks, but whether rising hot basaltic magma could cause enough heating and melting of continental crust to create a granite batholith is unclear.

Geochemical and other evidence suggests that no single simple model accounts for all granitic batholiths. The mechanism(s) involved vary from batholith to batholith, and sometimes from pluton to pluton within one batholith. We return to the question of the origin of batholiths in connection with the subject of the growth of continents (chapter 12).

Summary

Igneous rocks are those crystallized from magma, a silicate melt. Most magmas are produced in the upper mantle, where temperatures are high enough, and pressures low enough, to allow melting. The amount of melting is increased, and the melting temperatures of minerals are reduced, by the presence of volatiles and by the presence of several different minerals in one rock. The principal source of elevated temperatures to cause melting is the geothermal gradient, the normal increase in temperature with depth in the earth. Once formed, a cooling magma normally crystallizes principal silicate minerals in a sequence predicted by Bowen's reaction series, beginning with olivine and calcic plagioclase and ending with quartz and potassium feldspar. The composition of a magma can be changed by various processes, including fractional crystallization, assimilation, and magma mixing. The grain size of an igneous rock is determined fundamentally by cooling rate: All other factors being equal, slower cooling means larger crystals. Very rapid cooling produces fine-grained or even glassy rocks. Melt composition and the presence or absence of dissolved volatiles also influence grain size. Typically, plutonic rocks, which crystallize at depth, are coarser-grained than volcanic rocks.

Igneous rocks are classified on the basis of grain size and chemical composition. For most compositions, there are plutonic and volcanic equivalents, with different names reflecting the depths of crystallization. The mafic volcanic rock basalt is the principal rock type of the sea floor; continental crust is mainly granitic or granodioritic. Intrusive igneous rocks form plutons that may be pipelike, sheetlike, or fairly equidimensional. Plutons are classified on the basis of their shape and whether they are concordant or discordant with respect to the country rock. The shape of a pluton is controlled by both the physical properties of the magma and those of the country rock. Plutons may exhibit such additional features as chilled margins, compositional layering, and flow textures. The origin of the large plutonic complexes called batholiths is problematic, for no single mechanism (partial melting of mantle, fractional crystallization of basalt, melting of continental crust) seems able to account readily for the production of the necessary volume of granitic or granodioritic magma.

Terms to Remember

andesite
aphanitic
assimilate
basalt
batholith
Bowen's reaction
  series
chilled margin
concordant
continuous
  reaction series
country rock
dike
diorite
discontinuous
  reaction series
discordant
felsic
fractional
  crystallization
gabbro
geothermal
  gradient
granite
groundmass
igneous

laccolith
lopolith
mafic
magma
magma mixing
neck
obsidian
partial melting
pegmatite
phaneritic
phenocrysts
pipe
pluton
plutonic
porphyry
rhyolite
silicic
sill
stock
thin section
ultramafic
vesicles
volcanic
wallrock
xenolith

Questions for Review

1. What is an igneous rock?
2. Explain briefly the effect of each of the following on the melting temperature of rock: (a) changes in pressure, (b) presence of water vapor.

3. What is Bowen's reaction series? How do the continuous and discontinuous branches differ?
4. Describe two ways in which magma composition can be modified.
5. Assimilation of felsic rocks by mafic magmas is more common than the reverse. Why?
6. How is the grain size of an igneous rock related to its cooling rate? What does a porphyritic texture indicate?
7. Plutonic rocks may be more readily identifiable in handsample than volcanic rocks. Why?
8. Most volcanic glasses are rhyolitic in composition. Compositional layering is more often observed in mafic than in felsic plutons. What property of a magma may have a bearing on both of these observations? Explain.
9. What is a discordant pluton? Give two examples.
10. What is a batholith? Suggest two ways in which the necessary granitic magma might be formed, and discuss any problem with each idea.

For Further Thought

1. Considering magma viscosity and density, would you expect vol-

canic rocks more often to be mafic or silicic in composition? Why?
2. The minerals quartz, albite, and orthoclase (a potassium feldspar) are sometimes called, collectively, the "residua system" in the study of igneous rocks. How do you suppose this name arose?

Suggestions for Further Reading

Barker, D. S. 1983. *Igneous rocks.* Englewood Cliffs, N.J.: Prentice-Hall.

Bowen, N. L. 1956. *The evolution of the igneous rocks.* New York: Dover.

Carmichael, I. S., F. J. Turner, and J. Verhoogen. 1974. *Igneous petrology.* New York: McGraw-Hill.

Cox, K. G., J. D. Bell, and R. J. Pankhurst. 1979. *The interpretation of igneous rocks.* London: Allen and Unwin.

Ehlers, E. G., and H. Blatt. 1982. *Petrology: Igneous, sedimentary, and metamorphic.* San Francisco: W. H. Freeman.

Ernst, W. G. 1969. *Earth materials.* Englewood Cliffs, N.J.: Prentice-Hall.

Hughes, C. J. 1982. *Igneous petrology.* New York: Elsevier.

Maaloe, S. 1985. *Principles of igneous petrology.* New York: Springer-Verlag.

MacKenzie, W. S., C. H. Donaldson, and C. Guilford. 1982. *Atlas of igneous rocks and their textures.* New York: Halstead Press.

Outline

Volcanoes and
Volcanic Activity

The immense summit crater of Haleakala, a dormant shield volcano on the island of Maui, Hawaii, is more than 10 km across and dotted with small cinder cones formed late in its eruptive history. The silversword plant, shown in foreground, grows only on Haleakala.

Introduction

We have already surveyed some aspects of the composition, texture, and classification of volcanic rocks (chapter 3). Here, we focus particularly on the often dramatic events accompanying their formation: volcanic eruptions. A **volcano** is a vent through which magma, ash, and gases erupt, or the structure built around the vent by such eruption. Volcanoes vary considerably in character, in the kinds of materials they emit, and in the extent to which they pose hazards. Differences in volcanic character are directly related to magma composition and often to the plate-tectonic setting in which the volcano forms.

Before 1980, most Americans probably regarded volcanoes as remote phenomena, of no immediate relevance to their own lives. Then Mount St. Helens exploded with a roar in a cloud of ash, and volcanoes suddenly and forcefully demanded Americans' attention (figure 4.1). More recently, Hawaiian and Aleutian volcanoes have continued to be a reminder of the great store of heat and molten rock beneath the earth's surface.

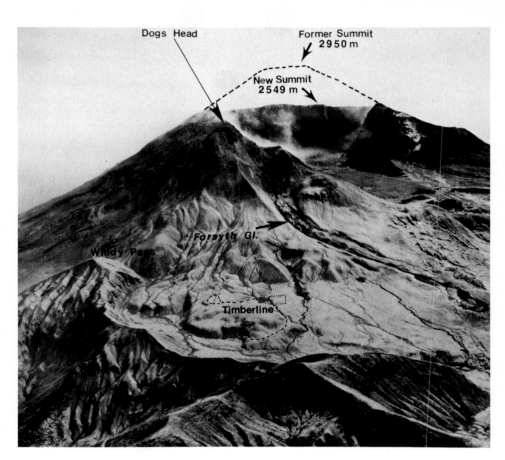

Figure 4.1 Aftermath of explosion of Mount St. Helens, 18 May 1980.
Photograph by A..Post, USGS Photo Library, Denver, Colorado

Kinds of Volcanic Activity

Much volcanic activity is a consequence of the eruption of **lava**—magma that reaches the earth's surface. Lava is only one product of volcanic eruptions, however. Nor are all volcanic rocks produced by individual volcanic cones. In chapter 3, a volcanic rock was defined as one formed from a magma—a silicate melt—at or close to the earth's surface. By this definition, the largest volume of volcanic rock is actually produced at the seafloor spreading ridges, mentioned in chapter 1 and described more fully in chapters 9 and 13. Because it is out of sight under the oceans, it is an often-forgotten form of volcanism.

Fissure Eruptions

The outpouring of magma at spreading ridges is an example of **fissure eruption,** the eruption of magma out of a

Figure 4.2 Schematic diagram of a fissure eruption.

long crack in the lithosphere, rather than from a single pipe or vent (figure 4.2). There are also examples of fissure eruptions on the continents, in which many layers of lava are erupted in succession. One example in·the United States is the Columbia Plateau, an area of about 50,000 square kilometers (20,000 square miles) in Washington, Oregon, and Idaho, covered by layer upon layer of basalt, piled up over 1½ kilometers deep in places (figure 4.3). This area may represent the ancient beginning of a continental rift that ultimately stopped spreading apart; it is

no longer active. Even larger examples of such *flood basalts* are found in India and Brazil.

Shield Volcanoes

Basaltic lavas, relatively low in silica and high in iron and magnesium, are comparatively fluid, so they can flow very freely and far when erupted. Consequently, the kind of volcano they build is very flat and low in relation to its diameter. This low, shieldlike shape has led to the use of the term **shield volcano** for such a structure (figure 4.4).

Figure 4.3 Multiple lava flows from fissure eruption. Several separate lava flows piled one atop another can be seen in this photograph taken near Coulee City, Washington.
© B. F. Molnia/TERRAPHOTOGRAPHICS/BPS

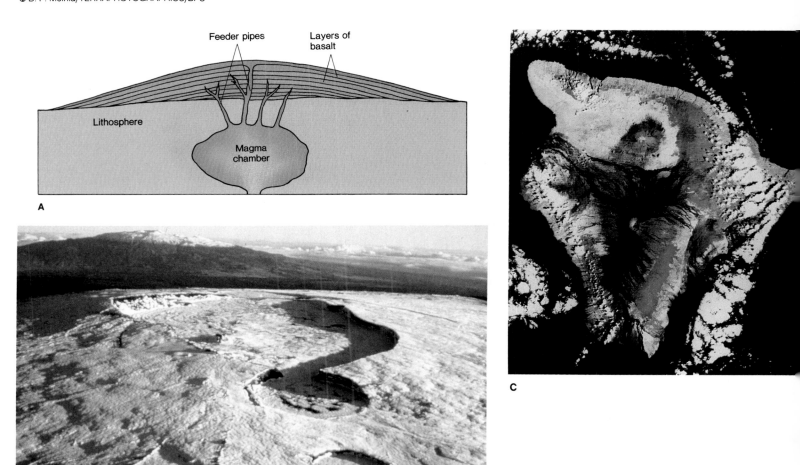

Figure 4.4 Shield volcanoes. (*A*) Schematic diagram of a shield volcano in cross section. (*B*) Mauna Loa, an example of a shield volcano, viewed from low altitudes. Note the flat shape. (*C*) Bird's-eye view of the island of Hawaii, taken by Landsat satellite, shows the island's volcanic character more clearly. The large peak is Mauna Loa; the smaller one is Mauna Kea.
(*B*) Photograph by D. W. Peterson, USGS Photo Library, Denver, Colorado. (*C*) © NASA.

Though the individual lava flows may be thin—perhaps less than a meter in thickness—the buildup of hundreds or thousands of flows over time can produce quite large volcanic structures. The Hawaiian Islands are all shield volcanoes. Mauna Loa, the largest peak on the still-active island of Hawaii, rises about 10 kilometers (6 miles) from the sea floor, with a base diameter of 100 kilometers. It stands taller above the sea floor than does Mount Everest above sea level. With their broad, flat shapes, the Hawaiian Islands do not necessarily look like volcanoes from sea level, but seen from above, their character is clear (figure 4.4C).

Even with limited variation in magma composition, there can be obvious variations in the appearance of the resultant lava flow. Especially fluid lavas that form a smooth, quenched, hardened skin as they cool develop a ropy appearance as they flow (figure 4.5A); this ropy lava is called **pahoehoe** (pronounced ''pa-hoy-hoy''). Other lavas flow less readily and produce jumbled, blocky flows (figure 4.5B); this material is called **aa** (''ah-ah'') and is reputed to have been so named from the pained cries of persons attempting to walk across its jagged surface barefoot. Some flows contract and fracture upon cooling to form polygonal columns, with the resultant structure described as **columnar jointing** (figure 4.5C). Underwater, the quenching of hot lava by seawater may produce **pillow basalts** (figure 4.5D).

Volcanic Domes

The less mafic, more silicic lavas, andesitic and rhyolitic in composition, tend to be more viscous and flow less readily. They ooze out at the surface

Figure 4.5 Types of lava flows and structures. (*A*) Pahoehoe, with a smooth, ropy surface: Hawaii. (*B*) Aa, rough and blocky: Lassen Park, California. (*C*) Columnar jointing: Devil's Postpile National Monument. (*D*) Pillow basalts on the sea floor.

(*D*) Photograph courtesy of W. R. Normark, USGS Photo Library, Denver, Colorado

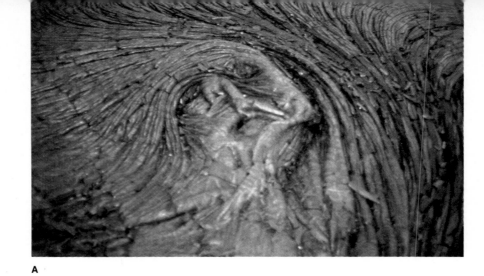

A

B

C

D

like thick toothpaste from an upright tube, piling up close to the volcanic vent, rather than spreading freely. The resulting structure is a more compact and steep-sided **volcanic dome** (figure 4.6A). Modern eruptions of Mount St. Helens are characterized by this kind of stiff, viscous lava, and a volcanic dome has formed in the crater left by the 1980 explosion (figure 4.6B). Such thick, slowly flowing lavas also seem to solidify and stop up the vent from which they are erupted before much material has emerged. Volcanic domes, then, tend to be relatively small in areal extent compared to shield volcanoes, although through repeated eruptions over time, such volcanoes can build quite high peaks.

Cinder Cones and Pyroclastic Eruptions

As noted previously, magmas consist not only of melted silicates, but also contain dissolved water and gases, which are trapped under great pressures while the magma is deep in the lithosphere. As the magma wells up toward the surface, the pressure on it is reduced, and the dissolved gases try to bubble out of it and escape. The effect is much like popping the cap off a soda bottle: The soda contains carbon dioxide gas under pressure, and when the pressure is released by removal of the cap, the gas comes bubbling out.

Sometimes, the built-up gas pressure in a rising magma is released suddenly and forcefully by an explosion that flings bits of magma and rock out of the volcano. The magma may freeze into solid pieces before falling to earth. The bits of violently erupted volcanic material are described collectively as **pyroclastics,** from the Greek words for "fire" (*pyros*) and "broken" (*klastos*). The most energetic pyroclastic eruptions are more typical of volcanoes with the more viscous andesitic or rhyolitic lavas because the thicker lavas tend to trap more gases. Gas usually escapes more easily and quietly from the fluid basaltic lavas, though even basaltic volcanoes may sometimes emit quantities of finer pyroclastics.

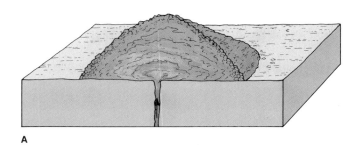

A

B

Figure 4.6   Volcanic domes. (*A*) Formation of a dome (schematic). (*B*) Dome built in the summit crater of Mount St. Helens.
(*B*) Photograph courtesy of R. E. Wallace, USGS Photo Library, Denver, Colorado

The fragments of pyroclastic materials can vary considerably in size. The very finest, which make up volcanic **ash,** range from a flourlike dust up to about 2 millimeters (0.1 inch) in diameter. **Cinders** are glassy, bubbly rock fragments that fall to the ground as solid pieces (figure 4.7A). The largest chunks, the volcanic **blocks,** can be the size of a house. Block-sized blobs of liquid lava may also be thrown from a volcano; these volcanic **bombs** commonly develop a streamlined shape as

they deform in flight before solidifying completely. A rock formed predominantly from ash-sized pyroclastic fragments is a **tuff** (figure 4.7B); the fragments may be so hot as they accumulate that they fuse together to form a *welded tuff.* A coarser rock containing large, angular blocks is a **volcanic breccia.** When pyroclastics fall close to the vent from which they were thrown, they may pile up into a very symmetric cone-shaped heap known as a **cinder cone** (figure 4.8).

A

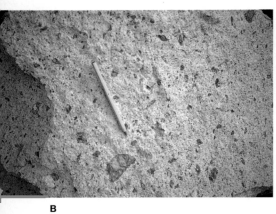

B

Figure 4.7 Pyroclastics and rock formed from pyroclastic material. (*A*) Cinders at Sunset Crater, Arizona. (*B*) The Bandelier Tuff, New Mexico.

A

B

Figure 4.8 Cinder cones. (*A*) Night eruption of Paricutín shows ejection of hot pyroclastics piling up in a cinder cone. (*B*) The resultant cinder cone at Paricutín.

(*A*) Photograph by R. E.Wilcox, USGS Photo Library, Denver, Colorado; (*B*) Photograph by K. Segerstrom, USGS Photo Library, Denver, Colorado.

CHAPTER 5

Weathering, Erosion, and Soil

Sediments and sedimentary rocks may erode
readily; here, gullying in the Chinle Formation,
Paria Canyon Wilderness, Utah.
© Wm. C. Brown Communications, Inc./Doug Sherman,
photographer

Introduction

At the earth's surface, rocks are constantly under attack—by gases in the air, by water and the chemicals dissolved in it, by wind-borne dust, ice, plant roots, and other agents. The resultant breakdown of rocks puts salts in the sea, produces the sediment of which sedimentary rocks are formed, and gives us the soil that is vital to plant growth and thus to our survival.

Broadly defined, **soil** is the surface accumulation of rock and mineral fragments (and, usually, some organic matter) that covers the underlying solid rock or sediment in most areas. Its formation, by definition, involves little or no transportation of material away from the site of formation. Once the products of weathering are transported and redeposited, they fall into the broader category of **sediment.** This chapter surveys principal weathering processes, controls on the composition of the resulting soil, and aspects of soil erosion and its minimization.

Weathering

The term **weathering** encompasses a variety of chemical, physical, and biological processes that act to break down rocks in place. The relative importance of different kinds of weathering processes is largely determined by climate. Climate, topography, the composition of the bedrock on which the soil is formed, and time determine a soil's final composition.

Mechanical Weathering

Mechanical weathering, also known as *physical weathering,* is the physical breakup of rocks, without changes in their composition. It requires that some physical force or stress be applied. Often, it is caused by water, as when water in cracks repeatedly freezes and expands, forcing rocks apart, then thaws and contracts or flows away. This process is termed **frost wedging.** Crystallizing salts in cracks may have the same wedging effect, as may plant roots

Figure 5.1 A tree root forcing its way into a crack contributes to mechanical weathering.

forcing their way into crevices (figure 5.1). *Abrasion* of rock surfaces by wind- or water-borne sediment is another form of mechanical weathering. Also, in extreme climates like those of deserts, where the contrast between day and night temperatures is very large, many cycles of the daily thermal expansion of rocks baked by sunlight and contraction as they cool at night might cause enough stress to break the rocks up, although the phenomenon has not been reproduced in the laboratory.

Removal of stress, too, can cause rocks to break up (figure 5.2). When a rock once deeply buried and under pressure is uplifted and the overlying rock eroded away, the rock is *unloaded.* It tends to expand and may fracture in concentric sheets in the process. As these sheetlike layers of rock flake off, the original rock mass is broken up by **exfoliation.** Granites, which would necessarily have crystallized at some depth initially, often weather by exfoliation when exposed at the surface.

Whatever the cause, the principal effect of mechanical weathering is the breakup of large chunks of rock into smaller ones. In the process, the total exposed surface area of the particles is increased (figure 5.3).

Chemical Weathering

Chemical weathering involves the breakdown of minerals by chemical reaction with water, with other chemicals dissolved in water, or with gases in the air. Minerals differ in the kinds of chemical reactions they undergo, as well as in how readily they weather chemically.

CHAPTER 6

Sediment and Sedimentary Rocks

Spectacular cross-bedding is exposed in a slot
canyon—Antelope Canyon, near Page, Arizona.
© David C. Schultz

Introduction

Sedimentary rocks represent the other end of the temperature spectrum from igneous rocks. They form at and near the earth's surface, from **sediments**—unconsolidated accumulations of rock, mineral grains, and organic matter that have been transported and deposited by wind, water, or ice. When the sediments become consolidated into a cohesive mass, they become sedimentary rock. The composition, texture, and other features of a sedimentary rock can provide clues about its origin, source materials, and the setting in which the sediment was deposited. Sedimentary rocks are widely exposed at the earth's surface, in part because they form close to the surface.

Classification of Sediments and Sedimentary Rocks

The ultimate source of sediment is the weathering of rock, described in chapter 5. Rocks exposed at the surface can be physically broken up into fragments; they can also undergo chemical reactions with air and water, reactions that produce new minerals and dissolve others. Chemicals thus dissolved and added to streams, lakes, and oceans can later be precipitated out of solution, with or without the aid of organisms. Sediments are classified, first, on the basis of the way in which they are formed, and second, on the basis of composition or particle size. The principal division is into *clastic* versus *chemical* sediments.

Clastic Sediments

Clastic sediments, which take their name from the Greek word *klastos*, meaning "broken," are composed of broken-up fragments of preexisting rocks and minerals. Individual fragments in a clastic sediment may be single mineral grains or bits of rock comprising several minerals.

A

B

Figure 6.1 Coarse-grained clastic sedimentary rocks. (*A*) Conglomerate—rounded fragments. Note the quarter for scale. (*B*) Breccia—angular fragments.
(*B*) © Wm. C. Brown Communications, Inc./Photograph by Doug Sherman

Clastic sediments, and the rocks formed from them, are named on the basis of the sizes of the fragments in them; see table 6.1. Gravels and boulders are the most coarse-grained clastic sediments. When they are transformed into rock, the corresponding rock is called either **conglomerate** or **breccia,** depending on whether the fragments are, respectively, well-rounded or angular (figure 6.1). The next finer sediments are sands, which, when consolidated, make **sandstone** (figure 6.2). Particles smaller than sand size are so fine that, in sediment, they usually don't feel gritty to the touch; individual grains are also indistinct or invisible to the naked eye. These very fine sediments are the silts and clays. Properly, the corresponding rocks are *siltstones*

Table 6.1 Simplified classification scheme for clastic sediments and rocks

Particle size range	Sediment	Rock
over 256 mm (10 in)	boulder	conglomerate (rounded fragments) or breccia (angular fragments)
2 to 256 mm (0.08 to 10 in)	gravel	conglomerate or breccia
1/16 to 2 mm (0.025 to 0.08 in)	sand	sandstone
1/256 to 1/16 mm (0.00025 to 0.025 in)	silt	siltstone*
less than 1/256 mm (less than 0.00015 in)	clay	claystone*

*Both siltstone and claystone are also known as mudstone, commonly called *shale* if the rock shows a tendency to split on parallel planes.

Figure 6.2 Examples of sandstone, a medium-grained clastic rock.
© Wm. C. Brown Communications, Inc./Photograph by Bob Coyle

Figure 6.3 Shale, a variety of mudstone. Note that individual grains are too fine to be seen with the naked eye.
© Wm. C. Brown Communications, Inc./Photograph by Bob Coyle

and *claystones.* However, since the grains cannot be seen with the naked eye, siltstones and claystones cannot readily be distinguished in hand-sample. They are often lumped together under the general heading of **mudstone,** a rock made from fine, muddy sediment. Mudstones commonly show a tendency to break roughly along planes corresponding to depositional layers, in which case they are given the name **shale** (figure 6.3).

None of the previous terms carries any implications about the composition of the fragments or even about whether the fragments are mineral grains or chunks of rock. Even the terms *clay* and *claystone* do not necessarily mean that the sediments consist of clay minerals, although clay minerals usually form only very fine crystals, and many mudstones and shales do indeed consist predominantly of clay minerals. (The abundance of clay minerals as products of chemical weathering accounts, in part, for the abundance of shales among sedimentary rocks.) However, clay-sized fragments of quartz, feldspar, obsidian, or other materials are also possible.

As already mentioned, the most coarse-grained clastic sediments—conglomerate and breccia—are distinguished on the basis of roundness or angularity of fragments. These properties are illustrated in figure 6.4. Angular fragments have sharp corners; the more well-rounded the fragments become, the smoother the corners. Sphericity is a measure of how equidimensional the grains are. Fragments

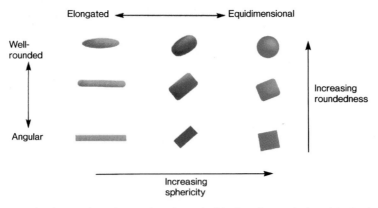

Figure 6.4 Roundedness versus grain shape. A grain can be elongated but well-rounded, or blocky (equidimensional) but angular.

Figure 6.5 Well-sorted and poorly sorted sediments. (*A*) Schematic: Grains are similar in size in well-sorted sediment. (*B*) Example of a well-sorted sandstone. (*C*) In this poorly sorted sediment, a wide range of grain sizes is present.
(*B*) © Wm. C. Brown Communications, Inc./Photograph by Bob Coyle

need not be spheroidal, like little balls, to be considered well-rounded; they must simply have smooth, rounded corners. With finer-grained sediments, distinctions are not made on the basis of roundedness. In the very finest sediments, roundedness cannot, after all, be determined with the naked eye because individual grains cannot be seen.

Another textural characteristic of clastic sediments and rocks is the degree of **sorting** they show. Sorting is a measure of the range in grain sizes present. A well-sorted sediment is one in which the grains are all similar in size; a poorly sorted sediment shows a wide variation in grain size (figure 6.5). The finer material filling in the pores between coarser grains in a poorly sorted sediment is termed the **matrix.** Sorting provides clues to the mode of sediment transport and environment of deposition. For example, sediments deposited directly by melting ice are typically poorly sorted; those deposited by wind or flowing water are often better sorted.

Chemical Sediments

Chemical sediments are those precipitated from solution. They may be either organic or inorganic in origin. The rocks formed from them are named, for the most part, on the basis of composition, not grain size. Several major types are summarized in table 6.2.

Volumetrically, the most important chemical sediment is **limestone,** a rock composed of carbonate minerals (figure 6.6). [Any rock consisting mostly of calcium and magnesium carbonate can be called limestone in the broad sense. However, the term is now generally restricted to rocks consisting of calcium carbonate, while a rock made of the calcium-magnesium carbonate mineral **dolomite** is more accurately called dolomite, or **dolostone.**] Carbonates may be precipitated from either fresh or salt waters. The oceans are an enormous reservoir of dissolved carbonate, and marine limestones hundreds of meters thick are known. Observations of modern sedimentation indicate that the majority of carbonate sediments are precipitated as calcium carbonate, whether directly from solution or, more commonly, from the deposition of calcium carbonate shells or skeletons of marine organisms. Most dolomite is believed to form from later chemical changes brought about through the addition of magnesium dissolved in pore waters circulating through the sediment, reacting with calcite to form dolomite.

When salt water is trapped in a shallow sea or lake and evaporates, dissolved minerals precipitate out as an **evaporite** deposit. Evaporites can contain a variety of minerals. Halite (sodium chloride) is an important one, as the oceans contain considerable dissolved salt. The resultant rock is called *rock salt* (figure 6.7). Other important minerals in evaporites are the calcium sulfates gypsum and anhydrite.

Table 6.2	Some chemical sedimentary rocks and constituent minerals
Rock type	**Composed of**
carbonate	
limestone	calcite, aragonite (both polymorphs of $CaCO_3$)
dolostone	dolomite ($CaMg(CO_3)_2$)
evaporite	variable—possibilities include halite (NaCl), gypsum (hydrated calcium sulfate, $CaSO_4 \cdot 2H_2O$), anhydrite ($CaSO_4$)
chert	silica (SiO_2)
banded iron formation	hematite, magnetite (iron oxides, Fe_2O_3, Fe_3O_4), usually with quartz or calcite

Figure 6.6 Limestones, examples of chemical sedimentary rocks.
© Wm. C. Brown Communications, Inc./Photograph by Bob Coyle

Under other conditions, dissolved silica (SiO_2) may precipitate out of solution, in a noncrystalline solid or even a watery gel. When solidified, such a deposit forms the rock *chert*. In some ancient sedimentary rocks, layers of iron oxides alternate with layers of chert or carbonate. These *banded ironstones* are important iron ores today (figure 6.8). Clays may sometimes precipitate from solution; some mudstones, therefore, are chemical sediments also.

Biological Contributions

Biogenic sediments—those produced by biological processes—do not really constitute a distinct class of sediments, except insofar as organisms are involved in their formation. Most are simply chemical sediments, organically

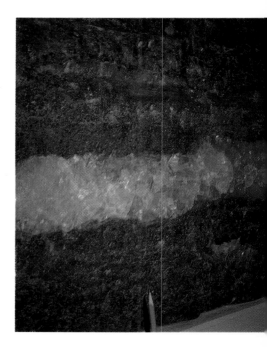

Figure 6.7 Rock salt, an evaporite.

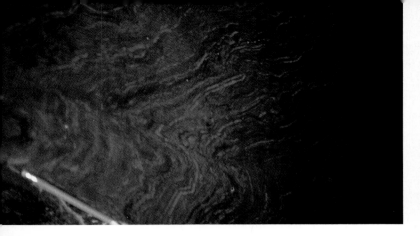

Figure 6.8   Banded iron formation, an iron-rich sedimentary rock.

Figure 6.9   A limestone rich in shell fragments.
© Wm. C. Brown Communications, Inc./Photograph by Doug Sherman

precipitated. A particular mineral is precipitated by an organism living in water, generally to build a shell or a skeleton; when the organism dies, and the soft body parts decay away, the mineral material remains. Coral reefs form carbonate rocks. In other cases, rocks are formed from sediments that consist of the remains of millions of microorganisms that have accumulated on the sea floor. Cherts can be formed from the silica skeletons of diatoms and radiolarians. Skeletons of calcareous (calcite-secreting) microorganisms can accumulate into a sediment from which limestone forms. In the modern oceans, such biogenic carbonate sediments may be more common than inorganically precipitated ones. Less commonly, coarser biological debris, such as carbonate shells, is transported and accumulated by wave action or currents in a deposit of carbonate sediment that produces a clastic limestone (figure 6.9). Our principal present fuel sources, the fossil fuels, are also biological sedimentary materials; see chapter 20.

Sedimentary Environments

The characteristics of a sedimentary rock are plainly determined in large measure by the depositional setting of the sediments. For clastic rocks, the nature of and distance from the source of the sediment are also factors. In this section, we survey briefly some of the common sedimentary environments and principal sediment characteristics of each.

Chemical Sedimentary Environments

Chemical sedimentation inevitably involves water, so all depositional environments of chemical sediments are under, or close to, water.

Shallow-water environments in warm waters account for most carbonate sedimentation. Unlike most chemicals, calcite dissolves more readily in cold water than in warm, or in other words, it is more readily precipitated from warmer waters. Off the coasts of some continents are broad, shallowly submerged shelves; in near-equatorial regions, the waters over these shelves are quite warm. In such areas, *carbonate platforms* may be built up by direct precipitation of carbonate muds or by the accumulation of shell or skeletal fragments of carbonate-secreting organisms that thrive under the same conditions. Reefs, likewise made of calcium carbonate, are common in this environment too (figure 6.10).

Restricted warm basins into which seawater can flow and from which it then evaporates provide the appropriate setting for deposition of evaporites. Such basins once existed over much of North America. Drilling into sedimentary rocks under the Mediterranean Sea likewise suggests that, at one time, water flow into and out of it was much more limited than it now is, so that seawater would flow in, be trapped, warmed, and evaporated, and leave a salt deposit behind. The thickness of many evaporite deposits is far greater than could be accounted for by taking a single basinful of seawater, even one as deep as the modern ocean, and drying it up. Apparently, thick evapo-

Figure 6.10   The extensive growth of Australia's Great Barrier Reef is partly due to favorable water temperatures (warm).

rite beds form through the cumulative effect of repeated cycles of basin filling and evaporation, or long periods of evaporation at rates exceeding the rate of influx of additional salt water.

Deep, cold basins on the sea floor may be sites of predominantly chemical sedimentation, if they are far enough removed from clastic input from the continents. Deep bottom waters are too cold for the preservation of calcium carbonate, but they are suitable for the preservation of silica. Biochemical silica-rich muds accumulate as the siliceous skeletons of diatoms and radiolarians settle to the sea floor.

Most voluminous chemical sedimentary rocks are produced from very saline waters. In certain circumstances, however, chemical precipitates from fresh water can produce unusual features; see box 6.1.

Outline

Metamorphic Rocks

Banding and folding in this gneiss show the effects of stress. This is an "augen gneiss," which derives its name from the German for "eye," referring to the eyelike appearance of rounded clumps of crystals in an elongated matrix of contrasting color.
© Wm. C. Brown Communications, Inc./Doug Sherman, photographer

Introduction

The term **metamorphism** is derived from Greek and Latin roots for "change of form." It has parallels in other sciences. For example, biologists use the term *metamorphosis* to describe collectively the complex changes by which larvae become butterflies and moths. The geologic *metamorphism* is likewise a process of change, but in rocks: change in the composition, mineralogy, texture, or structure of a rock by which it is transformed into a distinctly different rock. In this chapter, we survey some of the causes and consequences of metamorphism, describe situations that lead to metamorphism, and define some common metamorphic rock types.

Factors in Metamorphism

Rocks are changed when they are subjected to physical or chemical conditions different from those under which they formed. The two most common agents of metamorphism are heat and pressure. Each mineral is stable only within certain limits of pressure and temperature. Held for long enough under conditions in which it is unstable, a mineral changes to a more stable form. How long is long enough? That depends on the particular mineral and physical conditions.

Temperature in Metamorphism

Different minerals are stable over different temperature ranges. As temperatures increase, rocks are changed; minerals stable only at low temperatures break down or react to form higher-temperature minerals.

Metamorphic temperatures are limited at the low end by diagenesis, at the high end by melting. As noted in chapter 6, there is no sharp division between diagenesis and metamorphism—geologists disagree on how to define the boundary—but it generally corresponds to temperatures in the range of 100° to 200° C (200° to 400° F). The variability at the other end of the metamorphic temperature range is much greater because the temperature at which a rock melts depends

Figure 7.1   Metamorphic rocks exposed in a quarry in central Wisconsin. The banding and folding are the results of directed stress.

on many factors, including its composition, the prevailing pressure, and the presence or absence of fluids. Typical maximum metamorphic temperatures that most rocks can sustain without melting are in the 700° to 800° C (1,300° to 1,500° F) range, but dry mafic rocks rich in high-melting-temperature ferromagnesians can be heated to more than 1,000° C (1,850° F) without melting. Overall, the prevailing temperature influences not only the stability of various minerals but also the rates of chemical reactions, which generally proceed faster at high temperatures.

There are two principal sources of heat for metamorphism. One is the normal increase in temperature with depth already described, the *geothermal gradient*. This is a moderate effect in most places. Typical geothermal gradients are about 30° C per kilometer, or about 85° F per mile of depth. Local geothermal gradients can be increased by plutonism, emplacement of hot magma into the crust. Also, rocks very close to a pluton can be raised to near-magmatic temperatures directly. Plutonic activity, then, is a second source of heat for metamorphism.

Pressure in Metamorphism

Pressure takes two forms. Rocks in the earth are all subject to a **confining pressure,** an ambient pressure equal in all directions, imposed by the surrounding rocks, all of which are under

pressure from other rocks above. Confining pressure generally increases with depth. *Stress* is force applied to an object (such as a rock). In addition to confining pressure, rocks may or may not be subject to **directed stress,** which, as the name implies, is a force that is not uniformly applied in all directions.

The distinction between confining pressure and directed stress can be seen by analogy with a person standing on the earth's surface. The body experiences a confining pressure from the surrounding atmosphere (about 14.7 pounds per square inch of body surface at sea level), and atmospheric pressure acts equally all around the body. (If it did not, the body would be compressed in the directions from which higher pressures were exerted, stretched in others.) In addition, the feet are subjected to a vertical compressive stress applied by the downward pull of gravity on the mass of the body, and typically, the feet are somewhat flattened in response.

Directed stress in rocks frequently arises in connection with mountain-building processes, when rocks are squeezed, crumpled, and stretched. These *tectonic* processes are explored further in chapters 9 and 12. Directed stress can also develop on a smaller scale, as, for example, when magma is forcibly injected into cracks in rocks during igneous intrusion. The common result of directed stress is deformation (figure 7.1; see also box 7.1).

BOX 7.1

Stress and Strain under the Microscope

Strain or deformation on a gross scale is often visible as folds or other physical changes in outcrop (figure 7.1). Even when a rock does not obviously show such features, the effects of stress can often be seen in thin section. Compare figures 1

and 2. Figure 1 is a photomicrograph of a thin section of a granite, made up mostly of large, clear crystals of quartz (white to dark gray, featureless), coarse feldspar (flecked and striped), and biotite mica (greenish-brown blocks). Figure 2 is a

metamorphic rock (schist) of similar mineralogy, but the biotite laths are bent, and even the quartz grains have a mottled appearance due to small stress-induced irregularities in their crystal structures.

Figure 1   Thin section of granite. The dark, lathlike or blocky grains are biotite mica; the gray and white grains are quartz and potassium feldspar.

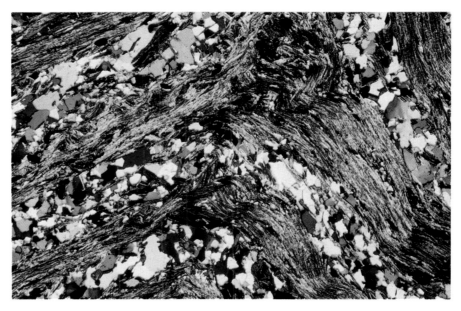

Figure 2   Thin section of biotite schist. Note the kinking or bending of biotite and the mottled appearance of the quartz grains due to stress.

Chemistry and the Role of Fluids in Metamorphism

A third factor that enters into some metamorphic situations is a change in chemical conditions in a rock. If rocks are fairly permeable, fluids flow readily through them, and either the fluids themselves, or one or more elements dissolved in them, may react with existing rock to produce new minerals. This can occur, for example, when warm fluid associated with magma migrates away from the silicate melt, carrying dissolved gases and other elements. The result is a change in bulk rock chemistry. **Hydrothermal** activity, involving the action of hot water seeping through fractures, is common in the country rock around igneous intrusions. Gases may also pass through or escape from permeable rocks.

The presence of fluid also facilitates mineralogical changes even if the rock as a whole behaves as a closed system, with no gain or loss of elements. Elements can migrate, or diffuse, through rock more readily in the presence of pore fluid than without it, so atoms can more readily rearrange themselves into new minerals in the presence of fluids. The hydroxide minerals and hydrous silicates, those containing (OH^-) groups, require the presence of some water (H_2O) even to form or to remain stable. Finally, as noted in chapter 3, wet rocks melt at lower temperatures than dry rocks, so the presence of fluid has some bearing on how high temperatures can get before the line between metamorphism and melting is crossed.

Effects of Metamorphism

The details of metamorphic changes depend both on the chemical composition and mineralogy of the *parent rock*, or starting material, and on the specific pressure, temperature, and other conditions of metamorphism, such as whether fluids or dissolved chemicals are entering or leaving the system. However, some generalizations are possible.

Table 7.1 Common types of metamorphic reactions

Reactions in silicate rocks

$(Mg,Fe)_7Si_8O_{22}(OH)_2$ = 7 $(Mg,Fe)SiO_3$ + SiO_2 + H_2O
amphibole, pyroxene, quartz, water

$(Mg,Fe)_3Si_2O_5(OH)_4$ + $KAlSi_3O_8$ = $K(Mg,Fe)_3AlSi_3O_{10}(OH)_2$ + 2 SiO_2 + H_2O
chlorite, potassium feldspar, biotite, quartz, water

3 $(Mg,Fe)_3(Al,Si)_2O_5(OH)_4$ + Fe_3O_4 + 6 SiO_2 = 3 $(Mg,Fe)_3(Fe,Al)_2Si_3O_{12}$ + 6 H_2O + ½ O_2
chlorite, magnetite, quartz, garnet, water, oxygen

$KFe_3AlSi_3O_{10}(OH)_2$ + ½ O_2 = $KAlSi_3O_8$ + Fe_3O_4 + H_2O
iron-rich biotite, oxygen, potassium feldspar, magnetite, water

Reactions in carbonate rocks

$CaCO_3$ + SiO_2 = $CaSiO_3$ + CO_2
calcite, quartz, wollastonite*, carbon dioxide

$CaMg(CO_3)_2$ + 2 SiO_2 = $CaMgSi_2O_6$ + 2 CO_2
dolomite, quartz, pyroxene, carbon dioxide

*Silicate similar to pyroxenes

Note: In all of the above reactions, the water involved is water vapor, carbon dioxide is a gas, and all other species are solids.

Many more reactions are possible, including various polymorphic transitions in which a low-temperature or low-pressure polymorph is converted to a higher-pressure or higher-temperature form (potassium feldspar, for example, has several polymorphs).

The reactions above should be regarded only as examples. Recall that certain mineral names (for example, "pyroxene" and "amphibole") encompass a number of specific mineral compositions that share common structural characteristics; so, for example, different compositions of pyroxenes are shown in the above equations.

Effects of Increased Temperature

As temperatures are increased, minerals that are stable at low temperatures react or break down to form high-temperature minerals. A common kind of reaction is one in which a hydrous mineral breaks down. Recall from Bowen's reaction series that the common hydrous silicates—amphiboles and micas—form at moderate to low temperatures. At higher temperatures, these minerals become unstable. For example, amphibole breaks down to form pyroxene and releases water; quartz is also produced, balancing the chemical equation:

$(Mg,Fe)_7Si_8O_{22}(OH)_2$ =
    amphibole
    7$(Mg,Fe)SiO_3$ + SiO_2 + H_2O
    pyroxene   quartz water

At even lower temperatures, the hydrous clay minerals in sedimentary rocks break down to form more stable micas; if temperatures are increased

further, the micas, in turn, break down (to feldspar and other minerals, plus water). Certain nonsilicate minerals may break down in analogous fashion. For instance, calcite and other carbonates may break down with release of carbon dioxide gas (CO_2). Examples of some possible metamorphic reactions are shown in table 7.1.

Effects of Increased Pressure

The effects of increased pressure depend, in part, on whether or not the stress is directed. In general, increased confining pressure favors denser minerals, those that pack more mass into less space. Garnet, for example, is a very common metamorphic mineral in moderate- to high-pressure situations. Iron-magnesium garnets are dense even by comparison with other ferromagnesian silicates: While the specific gravities of various pyroxenes and amphiboles are 3.2 to 3.9 and 2.0 to 3.4, respectively, ferromagnesian garnets

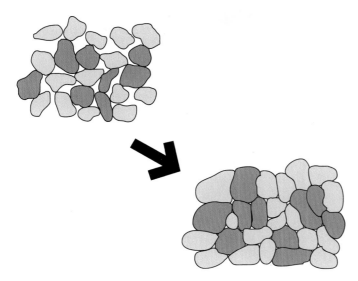

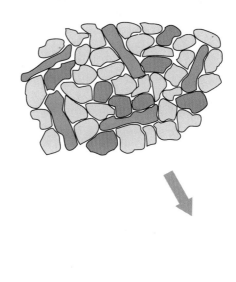

Figure 7.2 Schematic diagram of the effects of recrystallization during metamorphism. The grains become tightly interlocked, and porosity decreases.

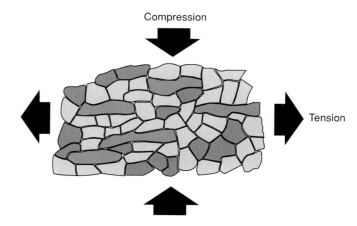

Figure 7.3 Development of foliation in the presence of directed stress. Elongated or platy minerals are reoriented, becoming aligned in parallel planes.

range from 3.8 to 4.3. Another effect of increased confining pressure that is particularly evident in metamorphosed sedimentary rocks is a tendency for the rock's texture to become much more compact. The minerals recrystallize and in so doing decrease the rock's porosity and increase its density (figure 7.2).

When directed stress is applied, the result is often a rock with some compositional or textural banding. This can take several forms. If there are already flat, platy minerals, such as clays and micas, in the rock, they tend to become aligned parallel to each other. *Tensional* stress tends to stretch the object to which it is applied; *compressive* stress tends to compress or squeeze the object. Platy minerals in a rock will tend to be drawn out parallel to a tensional stress, flattened perpendicular to a compressive stress, set parallel to the direction of a shear (figure 7.3). Even when the grains are too small to see with the naked eye, the rock often shows a tendency to split parallel to the aligned plates (rather than through or across the plates). This is **rock cleavage.** It is analogous to mineral cleavage, but rock cleavage is a tendency to break be-

tween planar mineral grains rather than between planes of atoms in a crystal. Rock cleavage is demonstrated by **slate,** which splits readily into flat slabs from which tiles and flagstones can be made (figure 7.4). In fact, rock cleavage developed in such a fine-grained rock is often called **slaty cleavage** because of this common example of the phenomenon.

As metamorphism intensifies at higher temperatures, crystals of various metamorphic minerals may grow progressively coarser (larger). Many metamorphic minerals have platelike or elongated, needle-shaped crystals

(micas and amphiboles, respectively, for instance). In the presence of directed stress, these crystals grow in similar preferred orientation. For example, flakes of mica tend to grow sideways to, rather than directly opposed to, a compressive stress (figure 7.5). The resultant rock texture, in which coarse-grained platy minerals show preferred orientation in parallel planes, is described as **schistosity** (figure 7.6). Slaty cleavage in fine-grained rocks and schistosity in coarser-grained ones are both examples of **foliation,** from the Latin for "leaf" (as in the parallel leaves, or pages, of a closed book).

Figure 7.4 Slate, showing characteristic slaty cleavage.
© Wm. C. Brown Communications, Inc./Photograph by Bob Coyle

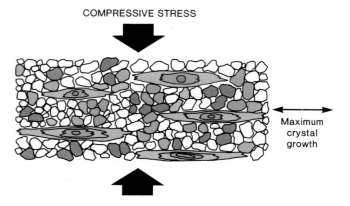

Figure 7.5 Growing micas grow most readily in the plane perpendicular to the applied compressive stress, leading to development of schistosity.

A

B

Figure 7.6 (*A*) Schistosity develops from the parallel alignment of mica flakes. (*B*) In outcrop, schistosity contributes to rock cleavage. Beaver Tail Light, Rhode Island.

Another texture sometimes regarded as a form of foliation is compositional, rather than textural, layering. In some metamorphic rocks, especially those subjected to strong metamorphism, recrystallizing minerals in the rock segregate into bands of differing composition or texture. Often, the result is a rock of striped appearance, with light-colored bands rich in quartz and feldspar alternating with dark-colored bands rich in ferromagnesian minerals (figure 7.7; see also figure 7.1). This mineralogical or compositional banding is *gneissic* (pronounced "nice-ic") texture. Although gneissic texture most commonly reflects compositional differences between layers, the term can also be applied to banded rocks with alternating schistose and equigranular layers.

On a megascopic scale, even rock fragments and fossils can be deformed or reoriented by directed stress. If one squeezes a tennis ball, it is flattened in the direction of compression and broadened in the plane perpendicular to it. A stretched piece of foam rubber is elongated in the direction of tension and thinned perpendicular to it. Rocks that have been stressed may likewise be obviously deformed; see figure 7.8.

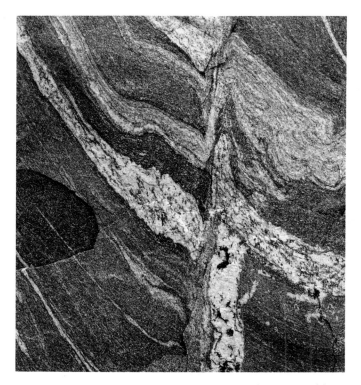

Figure 7.7 Gneiss, a compositionally banded metamorphic rock.

© Wm. C. Brown Communications, Inc./Photograph by Doug Sherman

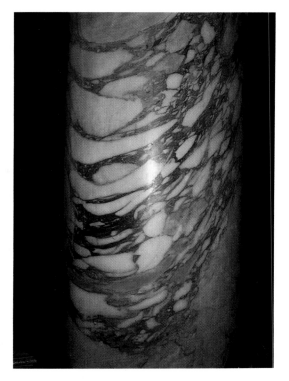

Figure 7.8 A metamorphosed marble breccia, showing stretched fragments.

Effects of Chemical Changes

It is difficult to generalize about the consequences of chemical changes during metamorphism. Briefly, addition of a particular element tends to stabilize minerals rich in that element and may cause more such minerals to form. For example, if migrating fluids carry dissolved potassium ions into a silicic rock, more potassium feldspars and micas might form. Such introduction of ions in fluids, called **metasomatism,** is not uncommon around plutons, from which fluids may escape into the surrounding rocks.

Conversely, loss of a specific chemical constituent has a destabilizing effect. For instance, if water is driven out of a permeable rock during metamorphism, any remaining hydrous minerals (clays, micas, amphiboles, and so forth) tend to break down into nonhydrous minerals (feldspars, pyroxenes, and others) more readily.

Many metamorphic minerals are not stable at surface conditions. Why don't these minerals all break down into low-temperature, low-pressure minerals when the rocks are exposed at the earth's surface? The main reason is lack of time. Both chemical reactions and structural changes like recrystallization proceed in solid materials quickly at high temperatures, much more slowly at low temperatures. Even over millions of years, some technically unstable minerals are preserved as a result of slow reaction rates. Thus, metamorphic rocks can be used to investigate pressure and temperature conditions that existed in the crust when the rocks were metamorphosed.

Types of Metamorphic Rocks

Metamorphic rocks are subdivided into *foliated* and *nonfoliated* rocks, on the basis of the presence or absence of preferred orientation of elongated or platy minerals. Further subdivisions within the foliated rocks are based largely on texture. Most unfoliated rocks are named on the basis of composition.

Foliated Rocks

Many of the common foliated rock types have names directly related to their characteristic textures. As noted previously, for example, slate (figure 7.4) is a metamorphic rock exhibiting slaty cleavage. Typically, the individual mineral grains in slate are too fine to be seen with the naked eye. The parent rock of a slate was generally a (fine-grained) shale. With progressive metamorphism of slate, the growing mica crystals may become coarse enough to begin to reflect light strongly, and the cleavage planes in the rock take on a shiny appearance in consequence. Such

Figure 7.9 Phyllite, showing glossy surface due to parallel mica flakes.
© Wm. C. Brown Communications, Inc./Photograph by Bob Coyle

Figure 7.10 Quartzite, showing a shiny surface of recrystallized grains.
© Wm. C. Brown Communications, Inc./Photograph by Bob Coyle

a rock is called **phyllite** (figure 7.9). The individual mica crystals in a phyllite may just be distinguishable with the unaided eye, but the rock is still basically fine-grained.

Further progressive metamorphism with the growth of coarser mica crystals in the presence of directed stress leads to the development of obvious schistosity, and the corresponding rock is called a **schist** (recall figure 7.6). A compositionally layered **gneiss**—a rock with gneissic texture, as shown in figure 7.7—may form either through still further metamorphism of a schist, with breakdown of some of the micas at higher pressures and temperatures, or during strong metamorphism of other rock types, such as granite.

Under extreme metamorphism, the lowest-melting-temperature minerals (generally quartz and potassium feldspar) may begin to melt, as the boundary between metamorphism and magmatism is reached. If this partly melted rock is cooled and fully solidified again, the resulting **migmatite,** or "mixed rock," has a mix of igneous and metamorphic characteristics. The rocks in figure 7.1 are migmatites.

The rock names of the foliated metamorphic rocks have few compositional implications. Parallel alignment of mica flakes creates slaty cleavage and schistosity, so slates and schists necessarily contain micas; even so, there are many different composi-

tions of micas, and other minerals may be present as well. The term *gneiss* indicates the existence of compositional layering without specifying anything at all about the minerals in the rock. Compositional terms can be added to the rock name to provide more information. For example, a rock might be described as a "garnet-biotite schist" if those two minerals are prominent in it; a "granitic gneiss" would be a metamorphic rock, showing gneissic texture, but having a mineralogic composition like that of granite (rich in quartz and feldspar).

Nonfoliated Rocks

Rocks that consist predominantly of equidimensional grains do not show foliation. If they consist mainly of a single mineral, they may be named on the basis of composition. When a quartz-rich sandstone is metamorphosed, for example, the quartz grains are recrystallized and the rock compacted into a denser product with tightly interlocking grains. The resulting metamorphic rock is called **quartzite** (figure 7.10). In hand specimen, it is not only denser than unmetamorphosed sandstone, but the recrystallized quartz grains in quartzite have a shinier or more glittery appearance than the original abraded grains in the sandstone. A metamorphosed limestone similarly recrystallized is

marble (figure 7.11; recall also figure 7.8). Usually, a marble is assumed to consist predominantly of calcite. If, in fact, it consists mainly of the mineral dolomite, it may be called a dolomitic marble for clarity.

An **amphibolite** is a metamorphic rock rich in amphiboles. Strictly speaking, amphibolites are not necessarily unfoliated rocks, for amphiboles commonly form in elongated or needlelike crystals that may take on a preferred orientation in the presence of directed stress. A textural term may then be inserted to indicate this: A foliated amphibolite, for example, is one in which the amphibole crystals lie in parallel planes.

Sometimes, a metamorphic rock neither shows a distinctive texture nor has a sufficiently simple composition to justify applying one of the previous terms. In such a case, if the nature of the parent rock can be recognized, the prefix *meta-* is simply added to the parent rock name to describe the present rock; for example, *metaconglomerate* or *metabasalt*.

Environments of Metamorphism

A variety of geologic settings and events lead to metamorphism. Most metamorphic processes can be subdivided into regional or contact metamorphism.

CHAPTER 8

Geologic Time

Millions of years of geologic history are
preserved, though imperfectly, in the layered
rocks flanking Heber Valley, near Wasatch
Mountain, Utah.
© David C. Schultz

Introduction

A student once wrote on an examination: "Geologic time is much slower and made up of very large units of time." This answer, while not quite the expected one, contains a large element of truth. Time passes no more slowly for geologists than for anyone else, but they do necessarily deal with the very long span of the earth's history. Consequently, they often work with very large units of time, millions or even billions of years long.

Much of the understanding of geologic processes, including the rates at which they occur and, therefore, the kinds of impacts they may have on human activities, is made possible through the development of various means of measuring ages and time spans in geologic systems. In this chapter, we explore some of these methods.

Relative Dating

Before any techniques for establishing numerical ages of rocks or geologic events were known, it was sometimes at least possible to place a sequence of events in the proper order. This is known as **relative dating.**

Arranging Events in Order

Among the earliest efforts to arrange geologic events in sequence were those of Nicolaus Steno. In 1669, he set forth two very basic principles that could be applied to sedimentary rocks. The first, the **Principle of Superposition,** pointed out that, in an undisturbed pile of sediments (or sedimentary rocks), those on the bottom were deposited first, followed in succession by the layers above them, ending with the youngest on top (figure 8.1). Today, this idea may seem rather obvious, but at the time, it represented a real step forward in thinking logically about rocks. Steno's second concept, the **Principle of Original Horizontality,**

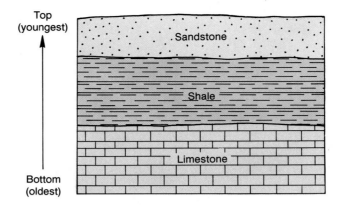

Figure 8.1 The Principle of Superposition. In an undisturbed sedimentary sequence, the rocks on the bottom were deposited first, and the depositional ages become younger higher in the pile.

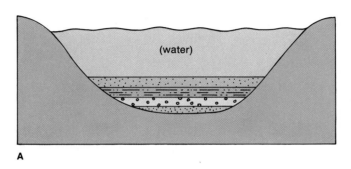

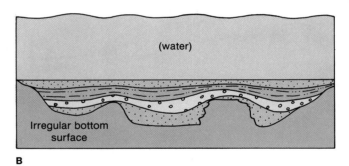

Figure 8.2 The Principle of Original Horizontality. (*A*) Sediments tend to be deposited in horizontal layers. (*B*) Even where the sediments are draped over an irregular surface, they tend toward the horizontal.

was based on the observation that sediments are commonly deposited in approximately horizontal, flat-lying layers (figure 8.2). Therefore, if one comes upon sedimentary rocks in which the layers are folded or dipping steeply, they must have been displaced or deformed after deposition and solidification into rock (figure 8.3).

Time-breaks in the sedimentary record also exist; they may or may not be easily recognized. An **unconformity** is a surface within a sedimentary sequence on which there has been a lack of sediment deposition, or where there has been active erosion, for some period of time. When the rocks are later examined, that time span will not be

Figure 8.3 These sedimentary rocks along the Oregon coast were tilted after deposition.

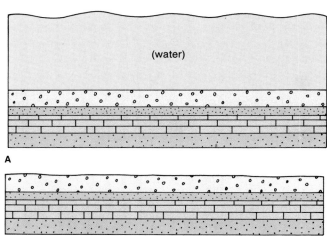

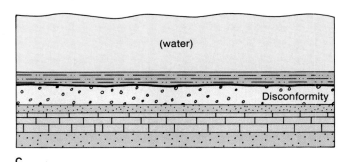

Figure 8.4 Development of disconformity: sediment deposition is interrupted for a period of time. (*A*) Sediment deposition in water. (*B*) Water now absent, deposition ceases. (*C*) Deposition resumes after some lapse of time; disconformity established.

represented in the record and, if the unconformity is erosional, some of the record once present will have been lost. (The term *nonconformity* is sometimes used to distinguish an unconformity at which erosion occurred during the gap in time.) The most difficult kind of unconformity to recognize is a **disconformity,** an unconformity at which the sedimentary rock layers above and below it are parallel (figure 8.4). Disconformities may be recognized by the presence of a weathered surface on the lower strata, or by juxtaposition of fossil forms known to be of very different ages. More obvious is the **angular unconformity,** at which the bedding planes in rock layers above and below the unconformity are not parallel (figure 8.5). The presence of an angular unconformity usually implies a significant period of erosion, to produce the new depositional surface for the younger rocks.

In later centuries, reasoning about the relative ages of rocks was extended to include igneous rocks in rock se-

quences. If an igneous dike or other pluton cuts across layers of sedimentary rocks, the sedimentary rocks must have been there first, the igneous rock introduced later. This is sometimes called the *Principle* (or *Law*) *of Cross-Cutting Relationships.* Often, there is a further clue to the correct sequence: The hotter igneous rock may have "baked," or metamorphosed, the country rock immediately adjacent to it, so again the igneous rock must have come second. (This is a particular help when the pluton is concordant.) If a pluton contains xenoliths, the rocks making up the xenoliths must predate the pluton (*Principle of Inclusion*).

Such geologic common sense can be applied to many rock associations. If a

strongly metamorphosed sedimentary rock is overlain by a completely unmetamorphosed one, for instance, the metamorphism must have occurred after the first sedimentary rock formed but before the sediments of the second were deposited. Quite complex sequences of geologic events can be unraveled by taking into account such simple underlying principles; see figure 8.6 for an example.

Correlation

Fossils play a role in the determination of relative ages, too. Around the year 1800, William Smith formulated the **Law of Faunal Succession.** The basic concept involved is that, through time,

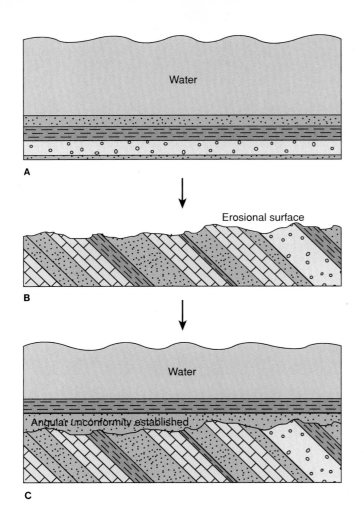

A

B

Erosional surface

C

Water

Angular unconformity established

D

Figure 8.5 Development of angular unconformity requires some deformation and erosion before sedimentation is resumed. (*A*) Deposition. (*B*) Rocks tilted, eroded. (*C*) Subsequent deposition. (*D*) Angular unconformity in the Grand Canyon.
© Wm. C. Brown Communications, Inc./Photograph by Doug Sherman

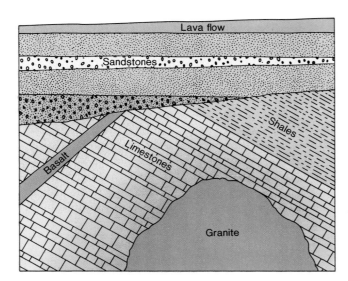

Figure 8.6 Deciphering a complex rock sequence. The limestones must be oldest (Principle of Superposition), followed by the shales. The granite pluton and basalt must both be younger than the limestone they crosscut. Note the metamorphosed zone around the granite. It is not possible to determine whether the igneous rocks predate or postdate the shales or whether the sedimentary rocks were tilted before or after the igneous rocks were emplaced. After the limestones and shales were tilted, they were eroded, and then the sandstones were deposited on top. Finally, the lava flow covered the entire sequence.

life forms change, old ones disappear from and new ones appear in the fossil record, but the same form is never exactly duplicated independently at two different times in history. This principle, in turn, implies that when one finds exactly the same type of fossil organism preserved in two rocks, even if the rocks are quite different compositionally and geographically widely separated, they should be the same age. Smith's law thus allowed age *correlation* between rock units exposed in different places (figure 8.7). A limitation on its usefulness, however, is that it can be applied only to rocks in which fossils are well preserved, which are almost exclusively sedimentary rocks. A theoretical basis for the Law of Faunal Succession was later provided by Charles Darwin's theories of evolution and natural selection.

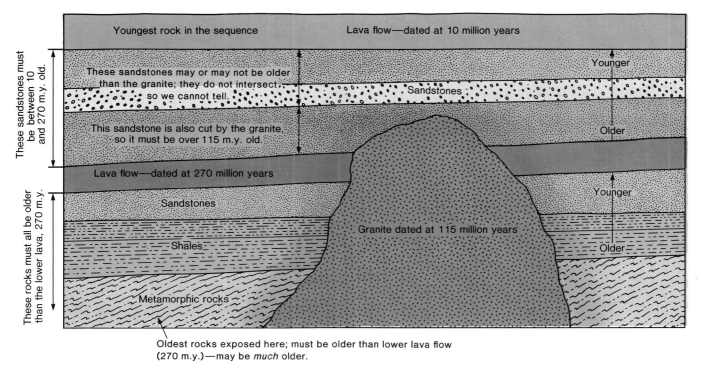

Figure 8.9 The ability to date some units of an outcrop allows the ages of units not directly dateable to be estimated.

With the advent of radiometric dating, numbers could be attached to the units and boundaries of the time scale. It became apparent that, indeed, geologic history spanned long periods of time and that the most detailed part of the scale, the **Phanerozoic** (Cambrian and later), was by far the shortest, comprising less than 15 percent of earth history.

Radiometric Dating and the Time Scale

Assigning very precise dates to the fine subdivisions of the Phanerozoic has proved difficult due to the nature of those subdivisions and of radiometric dating. As is apparent from previous discussion, the time of deposition of most sedimentary rocks cannot readily be dated. Very few fossils can be dated directly either, and in any case, fossilization may occur long after the organism is dead and buried. Yet the eras, periods, and epochs of the Phanerozoic were defined almost entirely on the basis of the sedimentary/fossil record. In practice, then, it was often necessary to approach the ages of the units by indirect dating and determination of age limits for sedimentary rocks

(where possible), as described earlier, and then making correlations with rocks in the type sections.

A further complication with respect to the finest subdivisions is that, like any measurements, most radiometric dates have some inherent uncertainty associated with them, arising out of geologic disturbance of the samples, laboratory and statistical uncertainties, and so on. In the best cases, these uncertainties are much less than 1 percent of the date determined, but some uncertainty always remains. Then, too, analytical methods and laboratory instrumentation have been constantly improving, and radiometric dates have been refined; redating of units dated previously has yielded new, more-precise ages that may differ slightly from the less-precise ages determined earlier.

The approximate time framework of the Phanerozoic was readily established, but the various technical limitations have caused persistent small uncertainties in the exact dates in question. The student should not be surprised, therefore, to find that texts published over the last few decades may differ somewhat in the exact ages shown.

The Precambrian has continued to pose something of a problem. A unit that spans 4 billion years of time seems to demand subdivision—but, in the absence of abundant fossils, on what basis? Moreover, Precambrian rocks are most often deformed and metamorphosed, so their original age and character are obscured. The first division into the *Archean* ("ancient") and *Proterozoic* ("pre-life") eons split the Precambrian into two nearly equal halves, but there was considerable disagreement as to how to define the boundary geologically. Although the time around 2.5 billion years ago was a time of widespread igneous activity, metamorphism, and mountain building, no single event of global impact could be identified to mark the boundary. Most common current usage simply takes the date of 2,500 million years as the Archean/Proterozoic boundary, without attempting to identify a specific geologic event to which this date corresponds. Universal agreement has not yet been reached on how the Archean and Proterozoic should be further subdivided.

Table 8.2 The Phanerozoic time scale

Era	Period	Epoch*	Date at boundary†	Distinctive life-forms
Cenozoic	Quaternary	Holocene		modern humans
			0.1	
		Pleistocene		Stone-Age humans
			2	
	Tertiary	Pliocene		
			5	
		Miocene		flowering plants common
			24	
		Oligocene		ancestral pigs, apes
			37	
		Eocene		ancestral horses, cattle
			58	
		Paleocene		
			66	
Mesozoic	Cretaceous			dinosaurs become extinct; flowering plants appear
			144	
	Jurassic			birds appear; mammals increase
			208	
	Triassic			dinosaurs, modern corals appear; first mammals
			245	
Paleozoic	Permian			rise of reptiles, amphibians
			286	
	Carboniferous‡			coal forests; first reptiles, winged insects
	Pennsylvanian			
			320	
	Mississippian			
			360	
	Devonian			first amphibians, trees, insects
			408	
	Silurian			first land plants, coral reefs
			438	
	Ordovician			first fish
			505	
	Cambrian			first widespread fossils, first vertebrates
			570	
(Precambrian)				scant invertebrate fossils—early shelled organisms, sponges, corals, worms

*Pre-Cenozoic periods are also subdivided into epochs, but there is little uniformity of nomenclature for these worldwide.

†Dates used from compilation of Geological Society of America for Decade of North American Geology; all in millions of years.

‡Not generally subdivided outside the United States.

How Old Is the Earth?— A Better Answer

Because the earth is geologically still very active, no rocks have been preserved unchanged since its formation. The age of the earth, therefore, cannot be determined directly, even by isotopic methods. However, evidence indicates that the earth, moon, and meteorites all formed at the same time, during the formation of the solar system. Fortunately, some of these other materials have had quite different subsequent histories. The moon is much smaller than the earth, cooled more rapidly after accretion, and has been geologically inactive for billions of years. The oldest of the samples returned by the manned Apollo lunar missions are approximately 4.6 billion years old.

Geologists also have numerous samples of *meteorites,* extraterrestrial fragments of rock and/or metal that have fallen to earth. Some of these have been disturbed after formation by collisions in space or by other processes, but the majority of meteorites yield ages in the 4.5 to 4.6 billion year range. Strong chemical similarities between the earth and meteorites support the idea that the earth and meteorites formed from the same materials (solar nebula), presumably at the same time.

On the basis of the foregoing and other evidence, the earth is inferred to have formed at about 4.55 billion years ago, and this date is typically taken as the beginning of Precambrian time. The oldest terrestrial rocks that have actually been accurately dated directly are close to 4 billion years old. The oldest rocks on each continent are generally between 3.6 and 3.9 billion years old.

Summary

Before the discovery of natural radio-activity, only relative age determinations for rocks and geologic events were possible. Field relationships and fossils were the principal tools used for this purpose. Rocks could be placed in relative age sequence or, with the aid of fossils, correlated from place to place.

Radiometric dating has made quantitative age measurements possible, although not all geologic materials can be so dated. Most of the commonly used isotopic dating methods work best for igneous and high-grade metamorphic rocks. It is rarely possible to determine the time of deposition of sedimentary rocks. This has led to some difficulties in the assignment of very precise ages to the units of the Phanerozoic time scale, which is subdivided largely using fossils and sedimentary rocks. The Phanerozoic, with its abundant life forms, represents only a relatively small part of the earth's history. The age of the earth is now estimated radiometrically at approximately 4.55 billion years.

Terms to Remember

angular unconformity	parent (nucleus)
Cenozoic	Phanerozoic
daughter (nucleus)	Precambrian
disconformity	Principle of Original Horizontality
era	
half-life	Principle of Superposition
Law of Faunal Succession	radioactive
Mesozoic	radiometric age
Paleozoic	relative dating
	unconformity

Questions for Review

1. Describe the significance of the Principle of Superposition and Principle of Original Horizontality to relative dating of sedimentary sequences.
2. What is the distinction between a disconformity and an angular unconformity? What do they have in common?
3. Explain two ways in which you might determine the relative ages of a pluton and surrounding sedimentary rocks.
4. How is the correlation of rock units made easier by the Law of Faunal Succession? What is a limitation on its use?
5. Why is it important to radiometric dating that radioactive isotopes have constant half-lives?
6. Describe any three requirements that must be satisfied for a radiometric decay scheme to be useful in dating geological materials.
7. It has proven somewhat difficult to establish radiometric dates for the units of the Phanerozoic time scale because the subdivisions were defined using sedimentary rocks. Explain. How do geologists address this problem?
8. When the geologic time scale was first established, the Precambrian was not subdivided. Why?
9. Why is it not possible to determine the age of the earth directly by radiometric methods? On what basis is its age estimated?

For Further Thought

1. Fossil forms used for correlation are sometimes termed *index fossils*. The best index fossils are those found widely distributed over the earth and that existed only for geologically short periods of time. Consider why these two criteria are important. What sorts of organisms might satisfy the first criterion especially well?
2. Each of the various radioactive isotopes has a distinct and unique half-life. What would be the impact on radiometric dating if all radioisotopes had the same half-life? If there were only one naturally occurring radioisotope?

Suggestions for Further Reading

Cloud, P. 1988. *Oasis in space: Earth history from the beginning.* New York: W. W. Norton.

Dott, R. H., Jr., and R. L. Batten. 1981. *Evolution of the earth.* 3d ed. New York: McGraw-Hill.

Eicher, D. L. 1968. *Geologic time.* 2d ed. Englewood Cliffs, N.J.: Prentice-Hall.

Faure, G. 1987. *Principles of isotope geology.* 2d ed. New York: John Wiley and Sons.

Harland, W. B. 1978. Geochronologic scales. American Association of Petroleum Geologists Studies in Geology #6, pp. 9–32.

Hurley, P. M. 1959. *How old is the earth?* New York: Doubleday. (This classic, written in the early days of radiometric dating of geological materials, is particularly aimed at the nonspecialist reader.)

Johns, R. B., ed. 1986. *Biological markers in the sedimentary record.* New York: Elsevier.

McLaren, D. J. 1978. Dating and correlation: A review. American Association of Petroleum Geologists Studies in Geology #6, pp. 1–7.

Moorbath, S. 1977. The oldest rocks and the growth of continents. *Scientific American* 236 (March): 92–104.

Nisbet, E. G. 1987. *The young earth: An introduction to Archean geology.* Boston: Allen and Unwin.

Windley, B. F. 1984. *The evolving continents.* 2d ed. New York: John Wiley and Sons.

CHAPTER 9

Outline

Obvious offset of stream channels across the
San Andreas Fault shows plate tectonics at
work.
Photograph by R. E. Wallace, USGS Photo Library,
Denver, Colorado

Plate Tectonics

Introduction

Several centuries ago, observers looking at world maps noticed the similarity in outline of the eastern coast of South America and the western coast of Africa (figure 9.1). Francis Bacon remarked upon the resemblance in the early seventeenth century. In 1855, Antonio Snider went so far as to publish a sketch showing how the two continents fit together, jigsaw-puzzle fashion. Such reconstruction gave rise to the bold suggestion that perhaps these continents *had* once been part of the same landmass, which had later broken up.

This concept of **continental drift**—the idea that individual continents could shift position on the globe—had an especially vocal champion in Alfred Wegener, a German meteorologist. He began to publish his ideas on the subject of continental drift in 1912 and continued to do so for nearly two decades. Several other prominent scientists found the idea plausible. However, most people, scientists and nonscientists alike, had difficulty visualizing how something as massive as a continent could possibly "drift" around on a solid earth or why it should do so. The majority of reputable scientists scoffed at the idea or, at best, politely ignored it.

Then, beginning in the 1960s, data of many different kinds began to accumulate that indicated that the continents have indeed moved. Continental drift turns out to be just one aspect of a broader theory known as **plate tectonics** that has evolved over the last two decades. **Tectonics** is the study of large-scale movement and deformation of the earth's outer layers. As noted in chapter 1, *plate tectonics* relates such deformation to the existence and the movement of rigid "plates" of rock over a weak or plastic layer in the upper mantle.

The Early Concept of Continental Drift

Alfred Wegener had based his proposition of continental drift on several lines of evidence, of which the fit of the

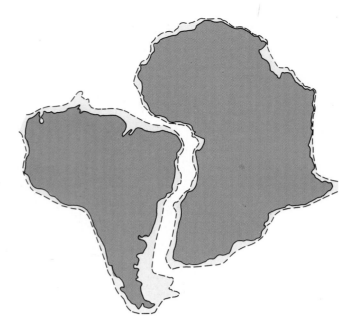

Figure 9.1 The jigsaw-puzzle fit of South America and Africa suggested that they might once have been joined together and were subsequently separated by continental drift. Blue shaded areas are the shallowly submerged continental shelves; dashed outlines, the outer limits of the shelves.

edges of the continents was only one. Another was that, when the continents were reassembled, similar rock types could be matched up between continents. Sedimentary rocks, in turn, contained further evidence in support of the idea.

Leaves of a distinctive fossil plant, *Glossopteris,* were found in southern Africa, Australia, South America, India, and even Antarctica. Certain dinosaur and other vertebrate remains were likewise found only over limited areas of several different continents. How did a given life form develop, identically and simultaneously, on different continents, now widely separated around the globe? Some paleontologists postulated long-distance transport of seeds and spores by wind or water to get over the intervening oceans. For the larger animals, they proposed the past existence of land bridges, now vanished, between continents. Wegener suggested instead that, at the time these organisms flourished, the continents had been united, and that they had split and drifted apart after the plant and animal remains had been entombed in the rocks.

Many factors determine a region's climate, but a dominant one is latitude. In general, equatorial regions tend to be warmest, polar regions coldest. Sedimentary rocks, formed at the earth's surface, often preserve evidence of the climatic conditions under which they formed. Fossil remains of plants known to thrive in moist heat imply a tropical climate; sandstones in which wind-blown desert dunes are preserved suggest dry conditions; some sediments can be identified as having been deposited by glaciers, as discussed in chapter 17.

In many cases, the ancient climate of a given region appears to have differed drastically from its present climate. For example, there is evidence of widespread past glaciation over much of India, in places where ice is unknown today. Moreover, such discrepancies cannot be accounted for simply in terms of *global* climatic changes, for the rock record does not show the same warming or cooling trends simultaneously on all continents. Wegener's interpretation was that rocks indicating cold climatic conditions were deposited when a continent was near a pole, and vice versa,

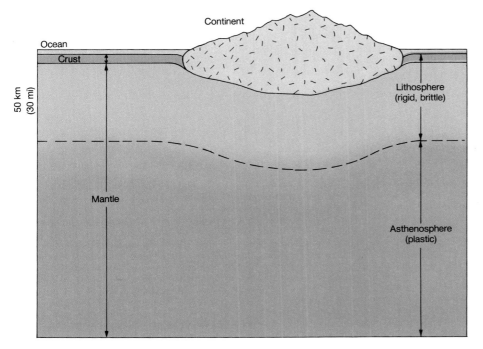

Figure 9.2 The outer zones of the earth (not to scale). The terms *crust* and *mantle* have compositional implications; *lithosphere* and *asthenosphere* describe physical properties. The lithosphere includes the crust and uppermost mantle. The asthenosphere lies entirely within the upper mantle.

and that marked changes in temperature regime were the result of the continents drifting from one latitude to another.

Wegener saw the continental blocks as plowing through the ocean floor, piling up mountains at their leading edges as they went. But there was no topographic evidence of disruption of the sea floor such as would be expected from this process. Nor could Wegener identify a convincing driving force to account for the movement. He died decades before his basic vision was validated.

Plates and Drifting Continents

One of the many obstacles to accepting the idea of continental drift was the difficulty of imagining solid continents moving over solid earth. However, as can be demonstrated by geophysical methods (see chapters 10 and 11), the earth is not completely solid from the surface to the center of the core. In fact, a plastic zone lies relatively close to the surface, so that, in a sense, a thin, solid, rigid skin of rock "floats" on a weaker, sometimes semisolid layer below (figure 9.2).

Lithosphere and Asthenosphere

The earth's crust and uppermost mantle are quite solid. The outer solid layer is termed the **lithosphere,** from the Greek word *lithos,* meaning "rock." The lithosphere varies in thickness from place to place on the earth. It is thinnest underneath the oceans, where it extends to a depth of about 50 kilometers (about 30 miles). The lithosphere under the continents is thicker, extending in places to depths of over 100 kilometers (60 miles).

The layer below the lithosphere, the **asthenosphere,** derives its name from the Greek word *asthenes,* meaning "without strength." This layer, which lies entirely within the upper mantle, extends to an average depth of about 500 kilometers (300 miles). Its lack of strength or rigidity may result, in part, from melting in the upper asthenosphere. This zone is certainly not all molten; at most, a small percentage of magma exists in otherwise solid rock. Such a small amount of melt would nevertheless cause the asthenosphere to behave plastically, rather than rigidly, under stress. Even where melting has not occurred, temperatures in the asthenosphere are so close to melting temperatures that, under such high-temperature, high-pressure conditions, the rocks behave plastically.

The presence of the asthenosphere makes the concept of continental drift more plausible. The continents need not drag across or plow through solid rock. Instead, they can be pictured as sliding over a softened or slightly mushy layer underneath.

Locating Breaks in the Lithosphere

The distribution of earthquakes and volcanoes on a map indicates that these phenomena are, for the most part, concentrated in belts or linear chains (see figures 4.11, 10.5). This suggests that the rigid shell of lithosphere is cracked and broken up into pieces, or plates. The volcanoes and earthquakes are concentrated at the boundaries of these lithospheric plates, where plates jostle or scrape against one another like ice floes on an arctic sea. About half a dozen very large lithospheric plates and many smaller ones have now been identified (figure 9.3).

Recognition of the existence of the asthenosphere made plate motions more plausible, but it did not *prove* that they had actually occurred. Likewise, the apparent existence of discrete plates of rigid lithosphere did not prove that those plates had ever moved. Many additional observations had to be gathered, and suggested explanations tested, before most scientists would accept the concept of plate tectonics. In time, it even became possible to document the rates and directions of plate movements.

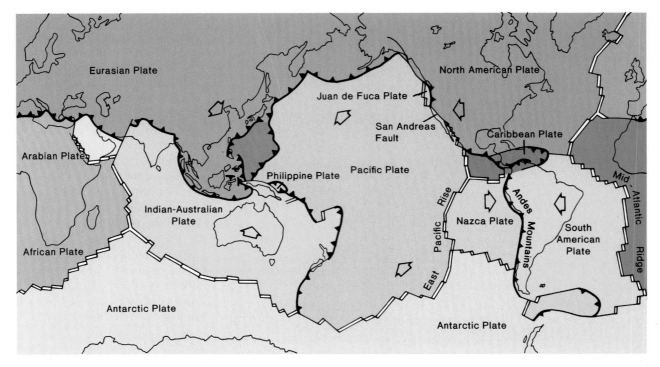

Figure 9.3 Principal world lithospheric plates.
Source: After W. Hamilton, U.S. Geological Survey.

Plate Movements— Accumulating Evidence

If South America and Africa really have moved apart, one might expect to see some evidence of the continents' passage on the sea floor between them. A topographic map of the floor of the Atlantic Ocean shows an obvious ridge running north-south about halfway between those continents (figure 9.4). This midocean ridge could be the seam from which the two continents have moved apart. Similar ridges are found on the floors of other oceans. As we shall see later in the chapter, it remained for scientists studying the ages and magnetic properties of seafloor rocks to recognize the significance of the ocean ridges to plate tectonics.

Magnetic Evidence from the Sea Floor

The earth possesses a magnetic field that can, to a first approximation, be described as similar to what would be expected from a huge bar magnet buried at the earth's center (figure 9.5). Magnetic lines of force run from the south magnetic pole to the north; the

magnetic poles lie close to the geographic (rotational) poles, though they are not identical. A compass needle aligns itself parallel to the lines of the magnetic field, north-south, and points to magnetic north. A dip needle, a magnetic needle suspended in such a way that it can pivot vertically, likewise dips along the trend of the magnetic field lines at that latitude (figure 9.6). Magnetic dip varies with latitude: The dip needle is horizontal near the equator, and its dip steepens to vertical at either pole.

Most iron-bearing minerals are at least weakly magnetic at surface temperatures. Each magnetic mineral has a **Curie temperature,** above which it loses its magnetic properties. The Curie temperature varies from mineral to mineral, but it is always below the mineral's melting temperature. A hot magma is therefore not magnetic, but as it cools and solidifies, and ferromagnesian silicates and other iron-bearing minerals crystallize in it, those magnetic minerals tend to align themselves parallel to the lines of force of the earth's magnetic field. They point to magnetic north and indicate a magnetic dip consistent with their magnetic latitude. They retain their internal

magnetic orientation unless they are heated again. This is the basis for the study of **paleomagnetism,** "fossil magnetism" in rocks. Paleomagnetism can also be detected in some iron-rich clastic sediments, in which the minerals have settled into alignment parallel to the earth's field, and in some metamorphic rocks, in which iron-rich minerals have recrystallized in similar orientation.

In the early 1900s, scientists investigating the direction of magnetization of a sequence of young volcanic rocks in France discovered that some of the earlier flows appeared to be magnetized in the opposite direction from the rest, their magnetic minerals pointing south instead of north. Confirmation of this discovery in many places around the world led to the suggestion, in the late 1920s, that the earth's magnetic field at some past time had "flipped," or reversed polarity, with north and south magnetic poles switching places.

Today, the phenomenon of magnetic reversals is well documented. Rocks crystallizing at times when the earth's field has been in its present orientation are said to be *normally magnetized;* rocks crystallizing when the field was oriented the opposite way are

Figure 9.4 The Mid-Atlantic Ridge, one prominent seafloor spreading ridge.

World Ocean Floor map by Bruce C. Heezen and Marie Tharp, 1977. Copyright © 1977 by Marie Tharp. Reproduced by permission of Marie Tharp, 1 Washington Ave., South Nyack, NY 10960.

described as *reversely magnetized.* Over the history of the earth, the magnetic field has reversed many times, but rocks of a given age show a consistent polarity. Through magnetic measurements and radiometric dating of the same rocks, geologists have been able to reconstruct the reversal history of the earth's magnetic field in detail. A portion of the recent reversal history is shown in figure 9.7. Just how or why the field reverses is still not clearly understood.

The ocean floor is made up largely of basalt, rich in ferromagnesian minerals. During the 1950s, the first large-scale surveys of the magnetic properties of the sea floor produced an entirely unexpected result. The floor of the ocean was found to consist of alternating "stripes" or bands of normally and reversely magnetized rocks, symmetrically arranged around the ocean ridges. At the time, geoscientists were baffled by this surprising discovery.

Then, in 1963, an elegant explanation was proposed by the team of F. J. Vine and D.H. Matthews and, independently, by L.W. Morley. A few years previously, geophysicist Professor Harry Hess of Princeton University had suggested the possibility of **seafloor spreading,** the idea being that the sea floor had split and spread away from the ridges. Seafloor spreading could account very simply for the magnetic stripes on the sea floor.

If the oceanic lithosphere splits and plates move apart, a rift in the lithosphere begins to open. But the result is not a 50-kilometer-deep crack. As rifting begins, some magma escapes from the asthenosphere. The magma rises, cools, and solidifies to form new basaltic rock, which becomes magnetized in the prevailing direction of the earth's magnetic field. As the plates continue to move apart, the new rock also splits and parts, making way for more magma to form still younger rock, and so on.

If, during the course of seafloor spreading, the polarity of the earth's magnetic field reverses, the rocks formed after the reversal are polarized oppositely from those formed before

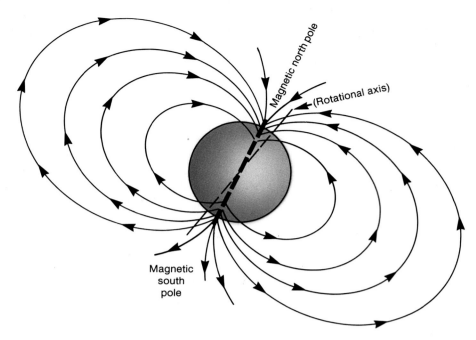

Figure 9.5 Lines of force of the earth's magnetic field.

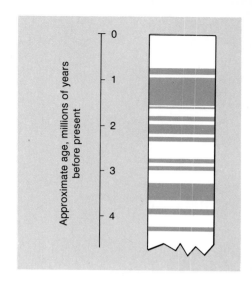

Figure 9.7 A portion of the reversal history of the earth's magnetic field for the recent geologic past.

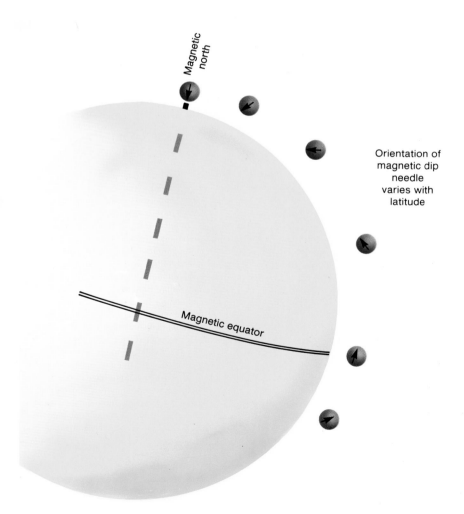

Orientation of magnetic dip needle varies with latitude

Figure 9.6 A dip needle varies in orientation with latitude, lying horizontally at the magnetic equator, pointing vertically at the poles.

it. The ocean floor is a continuous sequence of basalts formed over many tens of millions of years, during which time there have been dozens of polarity reversals. The basalts of the sea floor have acted as a sort of magnetic tape recorder throughout that time, preserving a record of polarity reversals in the alternating bands of normally and reversely magnetized rocks (figure 9.8).

Age of the Sea Floor

If any geologic model is correct, it should be possible to use that model to make predictions about other kinds of data before the corresponding measurements are made (the geologic version of the scientific method). The seafloor spreading model, in particular, implies that rocks of the sea floor should be younger close to the spreading ridges and progressively older farther away.

Specially designed research ships can sample sediment from the deep-sea floor and drill into the basalt beneath. Radiometric dating of the seafloor samples shows the predicted pattern. The rocks of the sea floor are youngest close to the ocean ridges and become progressively older the farther away they are from the ridges on either side

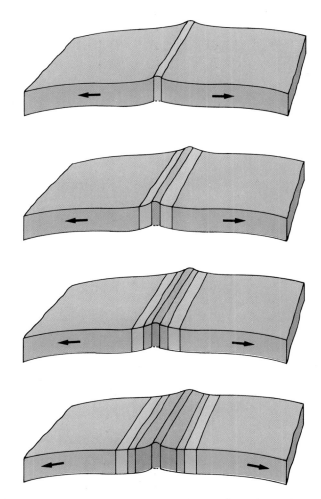

Figure 9.8 Formation of magnetic stripes on the sea floor. As each new piece of sea floor forms at the ridge, it becomes magnetized in a direction dependent on the orientation of the earth's field at that time. Past reversals of the field are reflected in alternating bands of normally and reversely magnetized rocks.

The discovery and construction of polar-wander curves was initially troublesome because there are good geophysical reasons to believe that the earth's magnetic poles should remain close to the geographic (rotational) poles, as they now are. Furthermore, the polar-wander curves for different continents do not match. Rocks of exactly the same age from two different continents may seem to point to two entirely different sets of magnetic poles!

This confusion can be eliminated, however, if it is assumed that the magnetic poles *have* always remained close to the geographic poles but that the *continents* have moved and rotated. The polar-wander curves then provide a way to map the directions in which the continents have moved through time, relative to the approximately stationary magnetic poles and relative to each other.

(see, for example, figure 9.9). Like the magnetic striping, the age pattern is symmetric across each ridge, again as one would predict. This is powerful confirmation of the seafloor-spreading hypothesis. The oldest rocks recovered from the sea floor, well away from the ridges, are about 200 million years old.

Polar-Wander Curves

Evidence for plate movements does not come only from the sea floor. For reasons outlined later in the chapter, much older rocks are preserved on the continents than beneath the oceans, so longer periods of earth history, and of paleomagnetic evidence, can be investigated through continental rocks.

Magnetized rocks of widely different ages on a single continent may point to very different apparent magnetic pole positions. The apparent ancient magnetic north and south poles may not simply be reversed but may be rotated from the present magnetic north-south. When the directions of magnetization of many rocks of various ages from a single continent are determined and plotted on a map, it appears that the magnetic poles have meandered far over the surface of the earth, if the position of the continent is assumed to have been fixed on the earth throughout time. The resulting curve, showing the apparent movement of the magnetic pole relative to one continent or region as a function of time, is the **polar-wander curve** for that landmass (see figure 9.10).

The Jigsaw Puzzle Refined

As mentioned earlier, the apparent similarity of the coastlines of Africa and South America triggered early speculation about the possibility of continental drift. Actually, the pieces of this jigsaw puzzle fit even better if we look not at coastlines but at the true edges of the continents—the outer edges of the continental shelves (dashed lines in figure 9.1). These represent the edges of the rifts that formed when a larger continental mass was split apart. Computers have been used to help determine the best physical fit of the pieces.

Continental reconstructions can be refined using details of continental geology—rock types, rock ages, evidence of glaciation, fossils, ore deposits, mountain ranges, and so on. If two now-separate continents were once part of the same landmass, then the geologic features presently found at the margin of one should have counterparts at the corresponding edge of the other. In short, the geology should match when the puzzle pieces are reassembled, as in the example shown in figure 9.11.

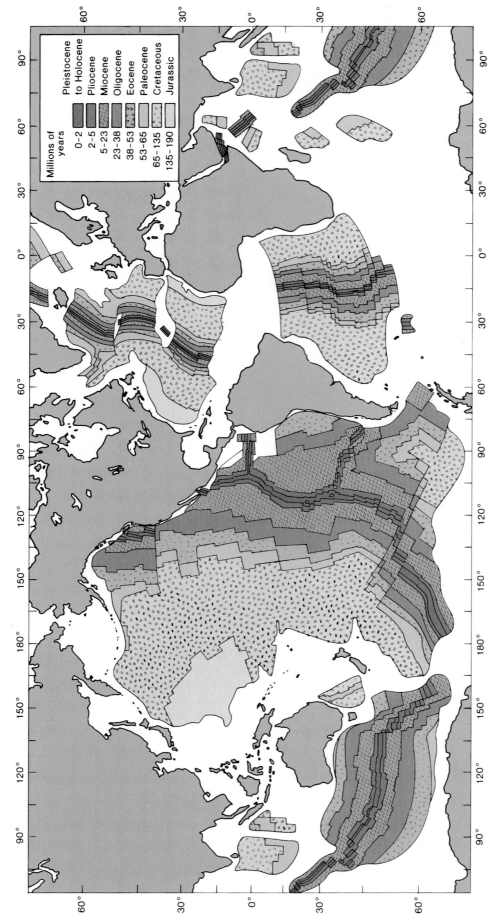

Figure 9.9 The pattern of seafloor ages on either side of the ridges reflects seafloor spreading activity. Younger rocks are closer to the ridge, and vice versa.

Source: Adapted from W. C. Pitman III, R. L. Larson, and E. M. Herron, Geological Society of America.

Millions of years

- Pleistocene to Holocene 0–2
- Pliocene 2–5
- Miocene 5–23
- Oligocene 23–38
- Eocene 38–53
- Paleocene 53–65
- Cretaceous 65–135
- Jurassic 135–190

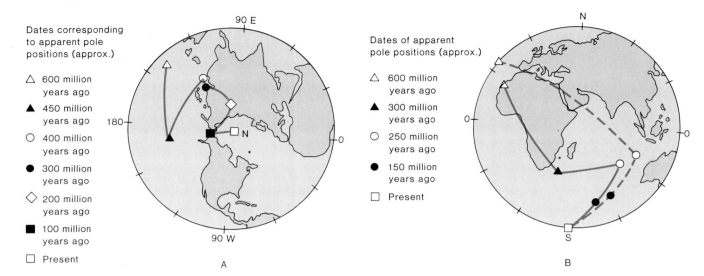

90 E
180
90 W
A

N
0
S
B

Figure 9.10   Examples of polar-wander curves. The apparent position of the magnetic pole relative to the continent is plotted for rocks of different ages and the data points connected to form the curve. The present positions of the continents are shown for reference. (A) Polar-wander curve for North America for the last 600 million years, as viewed from the North Pole. (B) Polar-wander curves for Africa (solid line) and Saudi Arabia (dashed line) suggest that these landmasses moved quite independently up until about 250 million years ago, when the polar-wander curves converged.

From M. W. McElhinny, *Paleomagnetism and Plate Tectonics.* Copyright © 1973 Cambridge University Press, New York, NY. Reprinted by permission.

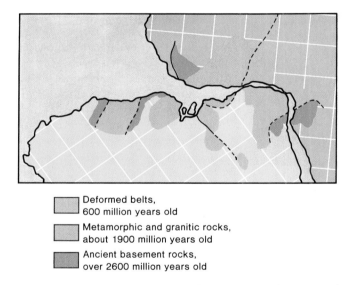

Deformed belts,
600 million years old

Metamorphic and granitic rocks,
about 1900 million years old

Ancient basement rocks,
over 2600 million years old

Figure 9.11   Geologic provinces of distinctive rock types and different ages can be correlated between western Africa and eastern South America.

From P. M. Hurley and J. R. Rand, *The Ocean Basins and Margins,* Vol. 1. Copyright © 1973 Plenum Publishing Corporation. Reprinted by permission.

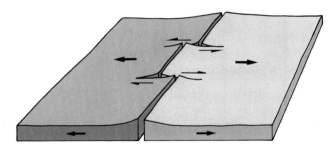

Figure 9.12   A spreading ridge, a divergent plate boundary. Arrows indicate direction of plate motions. Note transform faults between offset ridge segments (see figure 9.14 for detail in map view).

Types of Plate Boundaries

Different things happen at the boundaries between lithospheric plates, depending on the relative motions of the plates and on whether continental or oceanic lithosphere is at the edge of each plate.

Divergent Plate Boundaries

At a **divergent plate boundary,** such as a midocean spreading ridge, lithospheric plates move apart, magma wells up from the asthenosphere, and new lithosphere is created (figure 9.12). A great deal of volcanic activity thus occurs at spreading ridges. The pulling apart of the plates of lithosphere also results in earthquakes along these ridge plate boundaries.

Continents can be rifted apart, too (figure 9.13). This is less common, perhaps because continental lithosphere is so much thicker than oceanic lithosphere. In the early stages of continental rifting, volcanoes may erupt along the rift, or great flows of basaltic lava may pour out through fissures in the continent. As the continental crust is stretched and thinned, a shallow inland sea may inundate the rift zone. If the rifting continues, a new ocean basin floored with basalt will eventually form between the torn-apart pieces of continent. The Atlantic Ocean was created in this fashion. Iceland, sitting

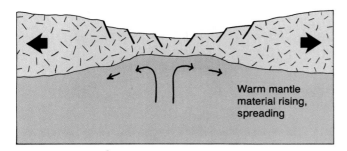

Figure 9.13 Continental rifting. The continental crust is thinned and fractured, and in time a new ocean basin is formed.

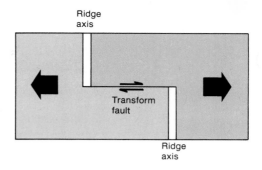

Figure 9.14 Transform fault between offset segments of a spreading ridge (map view). Arrows indicate the direction of plate movement. Along the transform fault between ridge segments, plates move in opposite directions.

From Charles C. Plummer and David McGeary, *Physical Geology*, 5th ed. Copyright © 1991 Wm. C. Brown Publishers, Dubuque, Iowa. All Rights Reserved. Reprinted by permission.

squarely on the Mid-Atlantic Ridge, is still being split by the parting of that rift. The Red Sea formed from a rift in continental lithosphere. Further rifting is also slowly ripping apart East Africa: The easternmost strip of the continent is being rifted apart from the rest, and another ocean may one day separate the resulting pieces. Through rifting, large plates are broken up into smaller ones.

Transform Boundaries

The actual structure of a seafloor spreading ridge is more complex than a single, straight crack. A close look at a midocean spreading ridge reveals that it is not a continuous break thousands of kilometers long (see figure 9.4). Rather, ridges consist of many short segments slightly offset from one another. The offsets are a special kind of fault, or break in the lithosphere, known as a **transform fault** (figure 9.14). The opposite sides of a transform fault belong to two different plates, and these are moving apart in opposite directions. As the plates scrape past each other, earthquakes occur along the transform fault between ridge segments. Beyond the spreading ridges to either side, both sides of the fracture belong to the same plate and move in the same direction.

The famous San Andreas Fault in California is an example of a transform fault that slices a continent sitting along a spreading ridge. The East Pacific Rise, a seafloor spreading ridge off the northwestern coast of North America, disappears under the edge of the continent, to reappear farther south in the Gulf of California (see figure 9.3). The San Andreas is the transform fault between these segments of spreading ridge. Most of North America is part of the North American Plate. The thin strip of California on the west side of the San Andreas Fault, however, is moving northwest with the Pacific Plate.

Convergent Plate Boundaries

At a **convergent plate boundary,** as the name indicates, plates are moving toward each other. Just what happens depends on what sort of lithosphere is at the leading edge of each plate. Continental lithosphere is relatively low in density, and buoyant, so it tends to float on the asthenosphere. Oceanic lithosphere is closer in density to the underlying asthenosphere, so it is more easily forced down into the asthenosphere as plates move together.

In a continent-continent collision, the two landmasses come together, crumple, and deform (figure 9.15). One may partially override the other, but neither sinks into the mantle, and a very large thickness of continent may result. Earthquakes are frequent during active collision as a result of the large stresses involved in the process. The extreme height of the Himalaya Mountains is attributed to just this sort of continent-continent collision, when India drifted northward over tens of millions of years and eventually collided with Asia.

More commonly, oceanic lithosphere is at the leading edge of one or both of the converging plates. One plate of oceanic lithosphere may be pushed under the other plate and descend into the asthenosphere. This type of plate boundary, where one plate is carried down below (*subducted* beneath) another, is called a **subduction zone** (figure 9.16). Topographically, there is commonly a depression (trench) marking the plate boundary on the sea floor.

The subduction zones of the world balance the seafloor equation. Because oceanic lithosphere is constantly being created at spreading ridges, an equal

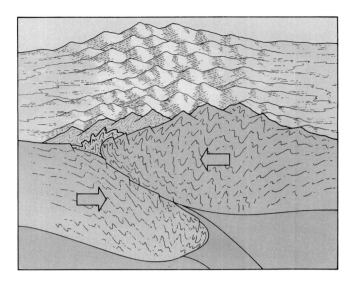

Figure 9.15 Continent-continent collision at a convergent boundary. Rocks are deformed and some lithospheric thickening occurs, but neither plate is subducted to any great extent.

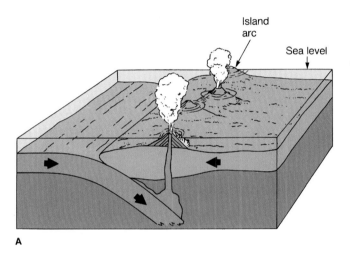

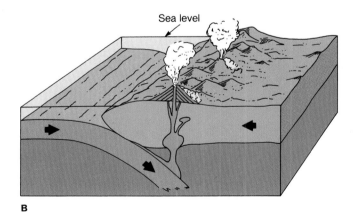

Figure 9.16 Subduction zone formed at a convergent boundary when a slab of oceanic lithosphere is forced into the mantle. (*A*) Ocean-ocean convergence: An island arc is formed by volcanic activity above the subducted slab. (*B*) Ocean-continent convergence: The magmatic activity accompanying subduction produces volcanic mountains on the continent.

amount of sea floor must be consumed in subduction zones. The subducted plate is heated by the hot asthenosphere and, in time, becomes hot enough to melt. Some of the melt rises to form volcanoes on the overriding plate. Some of the melt may eventually migrate to and rise again at a spreading ridge, to make new sea floor. In a sense, the oceanic lithosphere is constantly being recycled; entire plates may be consumed by subduction. This explains why no very ancient seafloor rocks are known. The buoyant continents are not so easily reworked.

Subduction zones are, geologically, very active places. Sediments eroded from the continents may accumulate in the trench formed by the down-going plate. Some of these sediments are caught in the fractured oceanic crust and carried down into the asthenosphere to be melted along with the sinking lithosphere. Volcanoes form where the melted material rises up through the overlying plate to the surface. Where the overriding plate also consists of oceanic lithosphere, the volcanoes may form a string of islands known as an **island arc** (figure 9.16A). Where continent overrides ocean, the volcanoes are built up as mountains on the continent (figure 9.16B). Mineral deposits may be formed in association with this magmatic activity. The great stresses involved in convergence and subduction give rise to large and frequent earthquakes. Parts of the world near or above subduction zones, and therefore prone to both volcanic and earthquake activity, include the Andes region of South America, western Central America, parts of the northwestern United States and Canada, the Aleutian Islands, China, Japan, and much of the rim of the Pacific Ocean basin; recall figure 9.3. The nature of volcanism and earthquake distribution at subduction zones are described in more detail in chapters 4 and 10 respectively.

How Far, How Fast, How Long, How Come?

Past Motions, Present Velocities

Rates and directions of plate movement can be determined in a variety of ways. As previously discussed, polar-wander curves from continental rocks can be used to determine the directions in which the continents have drifted. Seafloor spreading is another way of determining plate movement. The direction of seafloor spreading is usually obvious: away from the ridge. Rates of seafloor spreading can be found very simply by dating rocks at different distances from the spreading ridge and dividing the distance moved by the rock's age (the time it has taken to move that distance from the ridge at which it formed).

Still another way to monitor rates and directions of plate movement is by using mantle **hot spots.** These are isolated areas of volcanic activity, usually not associated with plate boundaries. Geologists believe that these volcanoes reflect unusual mantle conditions and more extensive melting beneath the hot spots. If we assume that mantle hot spots remain fixed in position while the lithospheric plates move over them, the result should be a string of volcanoes of differing ages, with the youngest closest to the hot spot.

A good example can be seen in the north Pacific Ocean (see figure 9.17). A relief map shows an L-shaped chain of volcanic islands and submerged volcanoes. When rocks from these volcanoes are dated radiometrically, they show a progression of ages, from about 75 million years at the northwestern end of the chain, to about 40 million years at the bend, through progressively younger islands to the still-active volcanoes of the island of Hawaii, at the eastern end of the Hawaiian Island group. The latter now sits nearly over the hot spot responsible for the whole chain. The age progression is reflected not only in the radiometric dates but also in topography and surface features: The farther west one goes in the chain, in general, the more extensively weathered and eroded the island, and the lower its relief above the surrounding ocean. Already, a new volcano and future island is forming east of Hawaii, as the Pacific Plate has continued to slide over the hot spot.

From the distances and age differences between pairs of points in the chain in figure 9.17, the rate of plate motion can be determined. For instance, Midway Island and Hawaii are about 2,700 kilometers apart. The volcanoes of Midway were active about 25 million years ago. Over the last 25 million years, then, the Pacific Plate has moved over the mantle hot spot at an average rate of (2,700 kilometers)/(25 million years), or about 11 centimeters per year. The orientation of the volcanic chain shows the direction of plate movement—most recently, west-northwest. The kink in the chain at about 40 million years ago indicates that the Pacific Plate changed direction at that time.

Looking at many such determinations from all over the world, geologists find that average rates of plate motion are 2 to 3 centimeters (about 1 inch) per year. In a few places, movement at rates up to about 18 centimeters per year is observed, and elsewhere, rates may be slower, but a few centimeters per year is typical. This seemingly trivial amount of motion does add up through geologic time. Movement of 2 centimeters per year for 100 million years means a shift of 2,000 kilometers, or about 1,250 miles! When the motions are extrapolated back into the past, the breakup of Pangaea, a single supercontinent (named from the Greek for "all lands") that existed 200 million years ago, can be reconstructed (figure 9.18).

Why Do Plates Move?

A driving force for plate tectonics has not been definitely identified. One widely accepted explanation, first suggested by Hess in connection with his seafloor spreading model, is that the plastic asthenosphere is slowly churning in large **convection cells** (figure 9.19). Hot mantle material rises at the spreading ridges; some escapes as magma to form new lithosphere, but most does not. The rest spreads out sideways beneath the lithosphere, slowly cooling in the process. As it flows outward, it drags the overlying lithosphere outward with it, thus continuing to open the ridges. When it cools, the flowing material becomes dense enough to sink back deeper into the mantle. This may be happening under subduction zones.

An alternative explanation for plate motions is that the weight of the dense, cold, down-going slab of lithosphere in the subduction zone pulls the rest of the trailing plate along with it, opening up the spreading ridges and reducing the pressure on the mantle below so magma can form and ooze upward. In this case, friction between the plate and asthenosphere below would help to drive convection as the plates move, rather than the reverse.

The full answer may be a combination of these mechanisms, and there may be contributions from other mechanisms not yet considered.

Antiquity of Plate Tectonics

How long plate-tectonic processes have been active is not entirely clear. The magnetic stripes characteristic of seafloor spreading are apparent over even the oldest, 200-million-year-old ocean floor. From continental rocks, apparent polar-wander curves going back more than a billion years can be reconstructed, although the relative scarcity of undisturbed ancient rocks makes such efforts more difficult for the earth's earliest history. It seems clear that the continents have been shifting in position over the earth's surface for at least 1 to 2 billion years, though not necessarily at the same rates or with exactly the same results as at present. For instance, if mantle convection is responsible for the motion, it may well have been more rapid in the past, when the earth's interior was hotter.

Figure 10.10 Building failure from 1985 earthquakes, Mexico City: Pancake-style collapse of fifteen-story reinforced-concrete structure.
Photograph by M. Celebi, courtesy of USGS Photo Library, Denver, Colorado

Figure 10.11 Trans-Alaska pipeline includes engineering features to allow "give" in response to stress.

The characteristics of the earthquakes in a particular region also must be taken into account. For example, severe earthquakes are generally followed by many **aftershocks,** earthquakes that are weaker than the principal tremor. The main shock usually causes the most damage, but when aftershocks are many and are nearly as strong as the main shock, they may also cause serious destruction. One day after the magnitude-8.1 main shock of the 1985 Mexico City earthquake, another major shock, with magnitude close to 7.3, leveled more buildings and deepened the rubble, hampering rescue efforts.

The duration of an earthquake also affects how well a building survives it. In reinforced concrete, ground shaking leads to formation of hairline cracks, which then widen and lengthen as shaking continues. A concrete building that can withstand a 1-minute main shock might collapse in an earthquake in which the main shock lasts 3 minutes. Many of the California building codes, used as models around the world, are designed for a 25-second main shock, but earthquakes can last ten times that long.

Finally, a major problem is that even the best building codes are typically applied only to new construction. Where a large city is located near a fault zone, thousands of vulnerable older buildings may already have been built in high-risk areas. The costs to redesign, rebuild, or even modify all of those buildings would be staggering.

A secondary hazard of earthquakes is *fire,* which may be more devastating than the results of ground movement. In the 1906 San Francisco earthquake, 70 percent of the damage was due to fire, not simple building failure. Fires start as fuel lines and tanks and power lines are broken, touching off flames and fueling them. At the same time, water lines are broken, leaving no way to fight the fires effectively. Putting numerous valves in all water and fuel pipeline systems helps to combat these problems because breaks in pipes can then be isolated before too much pressure or liquid is lost.

rock seem to suffer far less damage than those built on deep soil. Mexico City is underlain by thick layers of weak volcanic ash and clay; most smaller and older buildings lack the deep foundations needed to reach more stable sand layers at depth. This is one reason why damage from the 1985 Mexican earthquakes was so extensive in Mexico City, and why so many buildings completely collapsed. By contrast, Acapulco, which was much closer to the epicenter, suffered far less damage because it stands firmly on bedrock. The extent of damage in different parts of San Francisco as a consequence of the Loma Prieta earthquake was related to the underlying materials: Structures founded on Bay mud or artificial fill were damaged more than those on bedrock.

A

B

Figure 10.12 Landslides are often triggered by earthquakes. (*A*) Landslide in Turnagain Heights area, Anchorage, 1964. Close inspection reveals a number of houses amid the jumble of downdropped blocks in the foreground. (*B*) Landslide in the Santa Cruz Mountains from the Loma Prieta quake.
(*A*) Photograph courtesy of USGS Photo Library, Denver, Colorado. (*B*) Photograph by T. Holzer, USGS Open-File Report 89–687.

Ground Failure

Landslides can be a serious earthquake-related hazard in hilly areas since earthquakes are one of the possible triggering mechanisms of sliding of unstable slopes (figure 10.12). The best solution is not to build in such areas. Even if a whole region is hilly, detailed engineering studies of rock and soil properties and slope stability may make it possible to avoid the most dangerous sites. Visible evidence of past landslides is another indication of especially dangerous areas.

Ground shaking may cause a further problem in areas where the ground is very wet—in filled land, near the coast, or in places with a high water table. This problem is **liquefaction.** When wet soil is shaken by an earthquake, the soil particles may be jarred apart, allowing water to seep in between them. This greatly reduces the friction between soil particles that gives the soil strength. The ground then becomes like quicksand. When this happens, buildings can just topple over or partially sink into the liquefied soil; the soil has no strength to support them. The effects of liquefaction were dramatically illustrated after a major earthquake in Niigata, Japan, in 1964. One multistory apartment building tipped over to set-

Figure 10.13 Effects of soil liquefaction during an earthquake: Niigata, Japan, 1964. The buildings themselves were designed to be earthquake-resistant, and they toppled over intact.
Courtesy of the National Geophysical Data Center

tle at an angle of 30 degrees to the ground—while the structure remained intact! (See figure 10.13.)

In some areas prone to liquefaction, improved underground drainage systems may be installed to try to keep the soil drier, but little else can be done about this hazard, beyond avoiding areas at risk. Not all areas with wet soils are subject to liquefaction. The nature of the soil or fill plays a large role in the extent of the danger.

Figure 10.14 Boats washed into the heart of Kodiak, Alaska, by tsunami in 1964.
Photograph courtesy of National Geophysical Data Center

Tsunamis

Coastal areas, especially around the Pacific Ocean basin where so many large earthquakes occur, may also be vulnerable to **tsunamis.** These are seismic sea waves, sometimes improperly called "tidal waves," although they have nothing to do with tides. When an undersea or near-shore earthquake occurs, sudden movement of the sea floor may set up waves traveling away from that spot, like ripples on a pond caused by a dropped pebble.

Contrary to modern movie fiction, a tsunami is not seen as a huge breaker in the open ocean that topples ocean liners in one sweep. In the open sea, the tsunami is only an unusually broad swell on the water surface. Like all waves, tsunamis only develop into breakers as they approach shore and the undulating waters touch bottom. The breakers associated with tsunamis can easily be over 15 meters (45 feet) high in the case of larger earthquakes, and their effects correspondingly dramatic (figure 10.14). Several such breakers may crash over the coast in succession; between waves, the water may be pulled swiftly seaward, emp-

tying a harbor or bay and, perhaps, pulling unwary onlookers along. Tsunamis can travel very quickly—speeds of 1,000 kilometers per hour (600 miles per hour) are not uncommon—and tsunamis set off on one side of the Pacific may still cause noticeable effects on the other side of the ocean. A tsunami caused by a 1960 earthquake in Chile was still vigorous enough to make 7-meter-high breakers when it reached Hawaii some fifteen hours later, and twenty-five hours after the earthquake, the tsunami was detected in Japan.

Given the speeds at which tsunamis travel, little can be done to warn those near the earthquake epicenter, but people living some distance away can be warned in time to evacuate, saving lives, if not property. In 1948, two years after a devastating tsunami hit Hawaii, the U.S. Coast and Geodetic Survey established a Tsunami Early Warning System based in Hawaii. Whenever a major earthquake occurs in the Pacific region, sea-level data are collected from a series of monitoring stations around the Pacific. If a tsunami is detected, data on its source, speed, and estimated time of arrival can be relayed to

areas in danger, and people can be evacuated as necessary. Unfortunately, individuals' response to the warnings is variable. In 1985, when tsunami warnings were issued for parts of the coast of the Pacific Northwest, many people went to the shore to watch the expected waves, saying that, if danger threatened, they would run—clearly not realizing that tsunamis travel far too rapidly to outrace. Fortunately, tsunamis did not develop on that occasion.

Earthquake Control?

Since earthquakes are ultimately caused by forces strong enough to move continents, human efforts to stop earthquakes from occurring would be futile. However, moderation of some earthquakes' most severe effects may be possible.

As noted earlier, friction between rocks along existing faults may prevent movement until sufficient stress has built up to overcome the friction. Many of the worst earthquakes have occurred along sections of major faults that had for a time become **locked,** or immobile. Maps of the locations of earthquake epicenters along major faults show that there are sections with little or no seismic activity, while small or moderate earthquakes continue along other sections of the same fault zone. Such quiescent sections of otherwise-active fault zones are called **seismic gaps.** These areas may be sites of future serious earthquakes. On either side of a locked section, stresses are being released by creep and/or earthquakes. In the seismically quiet, locked sections, friction is apparently sufficient to prevent the fault from slipping, so the stresses are simply building up. The fear, of course, is that the stresses will build up so far that, when that locked section of fault finally does slip again, a very large earthquake will result. What is needed is a method of loosening locked faults gently, in a controlled way, rather than waiting for the sudden rupture of a large earthquake.

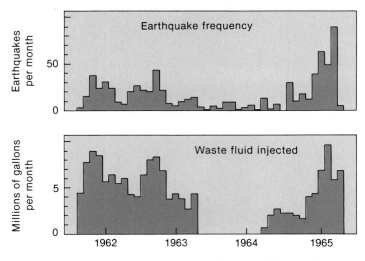

Figure 10.15 Correlation between waste disposal at the Rocky Mountain Arsenal (lower diagram) and frequency of earthquakes in the Denver area.

From David Evans, "Man-Made Earthquakes in Denver," *Geotimes*, Vol. 10, Number 9, pp. 11–18. Reprinted by permission.

In the mid-1960s, the city of Denver began to experience small earthquakes. They were not particularly damaging, but they were puzzling, since Denver had not previously been earthquake-prone. In time, a connection with an Army liquid-waste-disposal well at the nearby Rocky Mountain Arsenal was suggested. The frequency of the earthquakes was correlated with the quantities of liquid pumped into the well at different times (figure 10.15). The earthquake foci were also concentrated near the arsenal well. It seemed likely that the liquid, by increasing the fluid pressures in rocks along old faults, was decreasing the resistance to shearing and allowing the rocks to slip. Other observations and experiments around the world have supported the concept that fluids in fault zones may facilitate movement along a fault.

Such observations have prompted speculation by many scientists that **fluid injection** might be used along locked sections of major faults to allow the release of built-up stress. Unfortunately, geologists are presently far from sure

of the results to be expected from injecting fluid (probably water) along large, locked faults. There is no guarantee that only small earthquakes would be produced. Indeed, in an area where a fault had been locked for a long time, injecting fluid along that fault could lead to release of all the stress at once, in a major, damaging earthquake.

Possible casualties and damage are a tremendous concern, as are the legal and political consequences of a serious human-induced earthquake. Whether it is even possible to release large amounts of energy through small earthquakes only, considering the rate at which energy release increases with magnitude, is also questionable. The fluid-injection technique might be more safely used in major fault zones that have not been seismically quiet for long, where less stress has built up, to prevent that stress from continuing to build. Certainly, much more research is needed before this method can be applied safely and reliably on a large scale. It is, however, an intriguing possibility for the future.

Earthquake Prediction

Millions of people already live near major fault zones. For that reason, prediction of major earthquakes could result in many saved lives. Some progress has been made in this direction.

Earthquake Precursors

Earthquake prediction is based on the study of earthquake **precursor phenomena,** things that happen or rock properties that change prior to an earthquake. It may be possible to identify warning signs of an earthquake before it occurs. Many different, possibly-useful precursor phenomena are being studied. For example, the ground surface may be uplifted and tilted prior to an earthquake. P wave velocities in rocks near the fault (measured using artificially caused shocks, such as those from small explosions) drop and then rise before an earthquake. The same pattern is seen in the ratio of P wave to S wave velocity. Electrical resistivity (the resistance of rocks to the flow of electric current through them) increases, then decreases, before an earthquake. The time scale over which precursory changes occur varies: It may be on the order of weeks, months, years, or even decades. In a general way, the length of time over which precursory changes occur seems to be correlated with the size of the eventual earthquake: the longer the cycle, the larger the earthquake.

Other kinds of precursors may herald coming earthquakes. Changes in chemical properties of well water, including increases in radon concentrations, have been observed. Radon is a chemically inert, radioactive gas produced naturally in rocks by radioactive decay of trace quantities of uranium; it seems to seep more readily from rocks into groundwater prior to some earthquakes. Changes in water levels in wells have also been observed before earthquakes. The Chinese have had some success using anomalous animal behavior in prediction; it was a significant factor in the Haicheng prediction described in the next section. (Interestingly, they observed that animals that spend part or all of their time underground—snakes, for example—exhibited unusual behavior days before most surface-dwelling creatures, perhaps a not-unexpected observation.) In retrospect, there was some evidence of changes in periodicity of California geysers prior to the Loma Prieta earthquake in 1989.

These and other less well-documented or well-studied precursors have been considered in efforts to predict earthquakes. Such efforts are complicated by the fact that not all earthquakes show the same patterns of precursory events, aside from the problem that one must be monitoring a given fault zone in order to observe any precursory changes. A consequence of the present imperfect understanding of earthquakes is that, while there have been some spectacular successes in earthquake prediction, there have been equally conspicuous failures.

Current Status of Earthquake Prediction

In the People's Republic of China, some 10,000 scientists and technicians and 100,000 part-time amateur observers work on earthquake prediction. In February 1975, after months of smaller earthquakes, radon anomalies, and increases in ground tilt followed by a rapid increase in both tilt and microearthquake frequency, the scientists predicted an imminent earthquake near Haicheng in northeastern China. The government ordered several million people out of their homes, into the open. Nine and one-half hours later, a major earthquake struck, and many lives were saved because people were not crushed by collapsing buildings.

Over the next two years, the earthquake scientists successfully predicted four large earthquakes in the Hebei district. They also concluded that a major earthquake could be expected near T'ang Shan, about 150 kilometers southeast of Peking. In the latter case, however, they could only say that the event was likely to occur sometime during the following two months. When the earthquake—magnitude 8.2 with aftershocks up to magnitude 7.9—did occur, there was no immediate warning, no sudden change in precursor phenomena, and over 650,000 people died.

It may someday be possible to predict the timing and size of major earthquakes accurately enough that populated areas can be evacuated in an orderly way when serious earthquakes are imminent, thereby saving many lives. The property damage will still be inevitable as long as people persist in living and building in earthquake-prone areas. In any case, most seismologists feel that consistently reliable earthquake predictions are probably still more than a decade in the future. Only four nations—Japan, the former Soviet Union, the People's Republic of China, and the United States—have established government-sponsored earthquake prediction programs. As earthquake prediction becomes more commonplace, problems of appropriate response arise in turn; see box 10.1.

BOX 10.1

Seismic Risks and Public Response

Assuming that routine earthquake prediction becomes reality, some planners have begun to look beyond the scientific questions to possible social or legal complications of such predictions. For example, individuals have speculated that, if a major earthquake were predicted several years ahead for a particular area, property values would plummet because no one would want to live there. A counterargument to that view, perhaps, is San Francisco: Despite the widespread acceptance of the idea that another large earthquake like the 1906 earthquake will strike, despite the earthquakes of 1971 (San Fernando) and 1989 (Loma Prieta), the city continues to thrive. Nor, for that matter, are residents heeding the knowledge gained from previous earthquakes. The Marina district of San Francisco (figure 1) is built on fill. In 1857, Mark Twain observed that the worst damage from the earthquake of that year occurred in that part of the city. It suffered the worst damage in 1906. Workers picking up after the Loma Prieta quake unearthed rubble from 1906. The Marina district has now been restored following Loma Prieta . . .

The logistics of evacuating a large urban area on short notice if a near-term earthquake prediction is made is another concern. Many urban areas have hopelessly snarled traffic every rush hour; what happens if everyone in the city wants to leave at once? A frenzied evacuation might involve more casualties than the eventual earthquake. And what if a city is evacuated, people are inconvenienced or hurt, perhaps property is damaged by vandals and looters, and then the predicted earthquake never comes? Will the issuers of the warning be sued? Will future warnings be ignored?

The last question has been asked more often in the geologic community since the fall of 1990. A biologist-by-training who had become a business consultant made a "prediction" involving a major earthquake on the New Madrid fault zone, to occur about December 3, 1990. The prediction was not based on any precursors recognized by seismologists as significant; it related to the fact that on that date, the alignment of earth, sun, and moon in space would result in particularly strong tidal stresses on the earth. Uncritical media coverage of this "prediction", which was not endorsed by any official geoscience panel or body, resulted in near-hysteria in the New Madrid area. And the earthquake didn't happen. How will the public react to a scientific earthquake prediction in the future?

Figure 1 In San Francisco, earthquake damage is usually especially severe in the Marina district, where buildings are built on fill; here, damage from the 1989 Loma Prieta earthquake.
Photograph by M. Bennett, USGS Open-File Report 89–687.

CHAPTER 11

Outline

The Nature of the Earth's Interior

Mantle-derived basaltic magma gushing out at the surface makes lava fountains on Kilauea, Hawaii.
Hawaii Volcano Observatory photograph, USGS Photo Library, Denver, Colorado

Introduction

In several previous chapters, reference has been made to various aspects of the physical or chemical properties of the earth's interior. But how do geoscientists know so much about the interior? After all, earth's diameter is nearly 13,000 kilometers; the deepest wells, drilled to extract natural gas, penetrate only about 12 kilometers into the crust. Yet geoscientists discuss even the composition of the earth's core with some confidence. This chapter assembles the various kinds of evidence used to determine the physical and chemical makeup of the earth's interior.

Table 11.1 Composition of the earth's crust

Oxide	Continental crust	Oceanic crust	Overall average
SiO_2	59.2	48.0	58.1
Al_2O_3	15.4	15.2	15.1
FeO	7.5	10.7	7.4
MgO	4.3	7.7	4.4
CaO	6.0	12.2	6.8
K_2O	2.6	0.6	2.4
Na_2O	2.8	2.6	3.1
TiO_2	1.0	2.2	1.0

Note: All concentrations in weight percent; averages of several independent estimates.

Figure 11.1 Xenoliths of olivine-rich mantle material in basalt.

Chemical Constraints

Direct Sampling

Analysis of surface samples and drill cores reveals the compositions of a variety of crustal rocks of all types—igneous, metamorphic, and sedimentary. Where tectonic deformation has brought deep plutonic and high-grade metamorphic rocks closer to the surface, geoscientists have obtained some samples of lower-crustal rocks, though these exposures are few. Crustal rocks are extremely varied in composition, but the sampling is sufficiently thorough that there is general agreement on the average composition of the crust (table 11.1).

Deeper samples, from the upper mantle, are conveniently brought to the surface by volcanic rocks. Rising magmas formed by partial melting in the mantle may carry along with them fragments of still-unmelted mantle, or they may pick up pieces of mantle or crustal rocks as they ascend. These are *xenoliths*, first mentioned in chapter 3, and they provide firsthand clues to what lies beneath the surface (figure 11.1). The depths from which xenoliths come can frequently be deduced on the basis of the mineral assemblages in them. The same volcanic rock may contain xenoliths of several different subsurface rock units.

The deepest-source magmas originate at depths of approximately 200 kilometers. Thus, even the sampling of xenoliths only extends a few percent of the distance into the earth, and these deepest xenoliths are few. The xenoliths do, at least, include some upper mantle material. Further inferences about the composition of the upper mantle can be drawn from the compositions of the magmas produced from it, combined with laboratory experiments on melting and magmatic crystallization. The nature of a magma's parent mantle rocks can frequently be deduced from the composition of the partial melt.

Cosmic Abundances of the Elements

The compositions of stars can be determined from analyses of their spectra: Different wavelengths of light correspond to different elements. Most stars are quite similar in composition, at least in terms of the principal elements detected; 90 percent of them are compositionally similar to our sun. The sun's light spectrum can thus be analyzed to estimate the composition of most stars, which, in turn, encompass most of the mass of the universe. One limitation of this approach is that the presence and amounts of rarer elements are difficult to detect.

Fortunately, the major-element composition of the sun and of primitive meteorites are remarkably similar, excluding gases. This suggests that the relative amounts of the rarer elements in the same meteorites could be used to estimate the abundances of those elements in the rest of the solar system, and perhaps the universe.

In such ways, the so-called **cosmic abundance curve** of the elements can be assembled (figure 11.2). Relative abundances of most major elements in the earth's crust are also very similar to those shown. The principal discrepancies are in such elements as the inert gases and volatile elements (like hydrogen, oxygen, and carbon), and these may have failed to condense in the primitive earth as it formed from the solar nebula.

This, in turn, suggests that all of the nonvolatile elements should have condensed in the primitive earth. The estimated bulk composition of the earth thus determined is shown in table 11.2 along with an estimate of average crustal composition. If the crust is proportionately somewhat depleted in iron, nickel, and magnesium, for instance, perhaps these elements are more concentrated deeper in the earth. The abundance of ferromagnesians in upper-mantle samples is consistent with such reasoning. Conversely, one cannot realistically postulate that the earth's interior is made up of something relatively rare in the universe, like lead or gold.

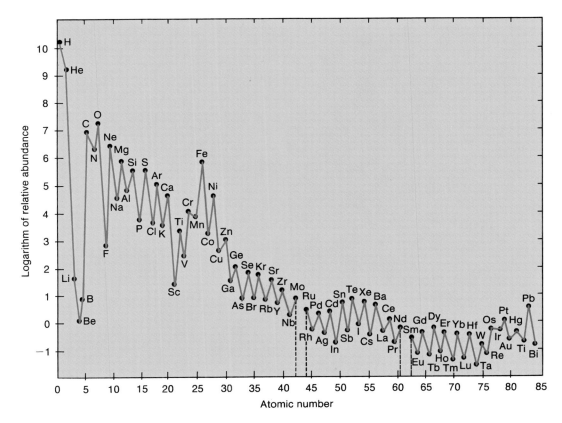

Figure 11.2 Portion of the so-called cosmic abundance curve of the elements. Note that the scale on the vertical axis is a logarithmic one. Abundances are reported per million (10^6) atoms of silicon.

Table 11.2	Approximate average composition of whole earth and of earth's crust	
Element	**Earth**	**Crust**
iron (Fe)	33.3	5.0
oxygen (O)	29.8	46.6
silicon (Si)	15.6	27.7
magnesium (Mg)	13.9	2.1
nickel (Ni)	2.0	0.01
calcium (Ca)	1.8	3.6
aluminum (Al)	1.5	8.1
sodium (Na)	0.2	2.8

Note: All concentrations in weight percent; averages of several independent determinations.

Physics and Geophysics

Bulk properties of the earth also restrict what geoscientists can propose for the earth's interior. The earth's gravitational field is a function of its mass. Mass and size together yield an average density for the earth of about 5.5 grams per cubic centimeter. This is about twice the density of continental crustal rocks and also significantly greater than the density of the ultramafic upper-mantle rocks of which geoscientists have samples. Even allowing for the great compression of the deep interior, it is still necessary to postulate the presence of materials in the interior that are considerably denser than any terrestrial materials actually sampled.

The Magnetic Field

The earth has a sizeable magnetic field, which could, so far as its orientation is concerned, be represented approximately by a giant bar magnet aligned nearly through the rotational poles. It might be tempting to postulate a large lump of solid iron, or perhaps iron-nickel alloy such as is found in some meteorites, in the middle of the earth. Iron is a relatively abundant element, it is strongly magnetic, and it is dense.

But putting a mass of solid iron in the interior does not account for the magnetic field as neatly as it might at first appear to do. Recall from chapter 9 that each magnetic material has a *Curie temperature,* above which it loses its magnetic properties. For iron, the Curie temperature is less than 800° C. Yet magmas from the upper mantle frequently have temperatures of 1,000° C or more; temperatures in the deep interior must be hotter still. Any solid iron deep within the earth, then, is well above its Curie temperature. The magnetic field has still to be accounted for, but the explanation must be more complex. Seismic evidence helps to solve the mystery.

Seismic Waves as Probes of the Earth's Interior

Seismic body waves provide a unique means of investigating the nature of the earth's interior below the depths from which samples can be obtained. Body waves from large earthquakes can be detected all over the earth, and their paths and travel times through the earth allow geoscientists to deduce such properties as the density and physical state of the materials at depth.

In a general way, body-wave velocities are proportional to the density of the material through which the body waves propagate: the denser the medium, the faster a given type of body wave travels. If body waves cross a boundary between two materials of distinctly different densities, they are **refracted,** or deflected in a different direction (figure 11.3). In a compositionally uniform material that increases smoothly in density with depth as a result of pressure, a body wave is continuously refracted, and the resulting path is curved (figure 11.4).

> **A**ll kinds of waves can be refracted. Light is similarly deflected on passing from air into water or a transparent mineral. The refraction of light waves can be observed by putting a hand into an aquarium or a pencil into a partially filled glass of water.

The Moho

One of the first internal features of the earth to be detected seismically was the crust/mantle boundary. It is marked by an increase in seismic-wave velocities in the mantle relative to the crust, which corresponds to a change in composition and consequent increase in density (confirmed by xenolith evidence). This boundary is the **Mohorovičić discontinuity,** often known as the **Moho** for short, named after the Yugoslavian seismologist who first reported on the seismic break by which it is identified.

The Moho is most easily recognized beneath the continents, where the greatest compositional contrasts between lower crust and upper mantle exist. Even under the continents, there is considerable lateral variation in seismic-wave velocities in crust and mantle, related to compositional variations. Commonly, the Moho in any

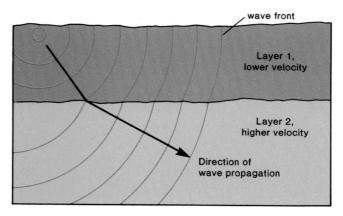

Figure 11.3 Refraction of seismic wave at a boundary between two layers of different density.

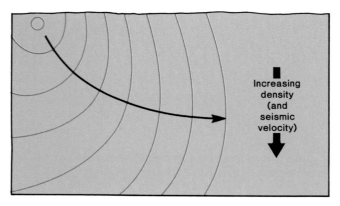

Figure 11.4 When density increases gradually with depth, continuous refraction in small increments produces a curved path.

spot is defined as the depth at which P-wave velocities first reach 7.8 kilometers per second or higher.

The higher seismic velocities in the upper mantle and the phenomenon of wave refraction together make possible the measurement of the thickness of the earth's crust. The principle is illustrated in figure 11.5. When an earthquake occurs, the seismic body waves traveling only through the crust reach stations near the site ahead of those following a longer, deeper path to the mantle and back up to the surface. But at some distance away, the higher seismic velocities of the mantle compensate for the greater distance traveled by waves following the deeper routes, and those waves traveling part of the way in the mantle arrive ahead of waves traveling entirely in the crust.

The effect can be visualized by analogy. Suppose that a person has some distance to travel and can either walk directly to the destination or take a (faster) bus, which requires walking a bit out of the way at either end of the trip. For very short trips, the direct walk is the faster way to go, since it avoids the detour to the bus. But for long trips, the time spent getting to and from the bus is more than compensated for by the time saved along most of the length of the route.

How long the trip must be before taking the bus becomes time-effective depends partly on how much faster the bus travels than the pedestrian does and partly on how far out of the way the bus route is. Likewise, how far apart earthquake and seismograph must be for the body waves going down to the

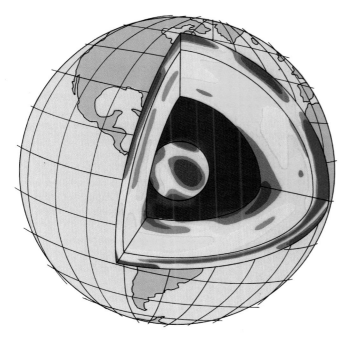

Figure 11.13 Cutaway view of seismic-velocity variations in the earth's interior. Blue (colder?) regions have above-average velocities, red (warmer?) below-average. Velocity variations in the upper mantle are for S waves; total variation may be up to 10%. In deeper regions, variations are for P-wave velocities, which vary laterally by less than 2%.

From A. N. Dziewonski, "Global Images of the Earth's Interior," in *Science* 236:37, April 3, 1987. Copyright © 1987 by the American Association for the Advancement of Science. Reprinted with permission.

Beyond the general increase of temperature with depth, there are lateral variations in temperature (and thus in density, viscosity, and other properties) in the earth's interior. Increasingly, these are being explored through **seismic tomography,** through which small variations in seismic velocity are used to map lateral variations in temperature. Use of this technique may lead to better understanding of mantle plumes, convection, and related phenomena. Figure 11.13 is an example of the kind of image of the interior that seismic tomography provides.

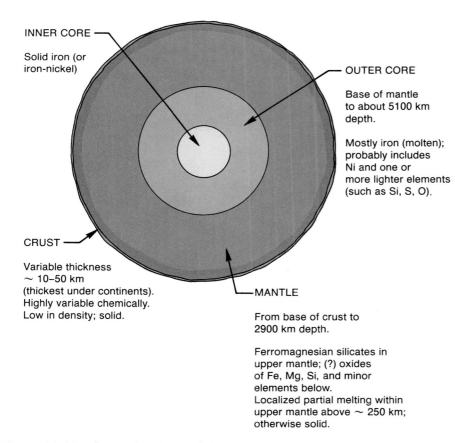

INNER CORE

Solid iron (or iron-nickel)

CRUST

Variable thickness ~ 10–50 km (thickest under continents). Highly variable chemically. Low in density; solid.

OUTER CORE

Base of mantle to about 5100 km depth.

Mostly iron (molten); probably includes Ni and one or more lighter elements (such as Si, S, O).

MANTLE

From base of crust to 2900 km depth.

Ferromagnesian silicates in upper mantle; (?) oxides of Fe, Mg, Si, and minor elements below. Localized partial melting within upper mantle above ~ 250 km; otherwise solid.

Figure 11.14 Composite picture of the earth's interior and its various zones.

Summary

Information of many kinds contributes to the determination of the physical state and chemical composition of the earth's interior, shown in figure 11.14. Chemical and mineralogical data for the crust and uppermost mantle are obtained by direct sampling and analysis, coupled with laboratory studies of magmas and melting. The bulk composition of the earth can be constrained using compositional data from meteorites and stars. Physical parameters (mass, density, and so forth) and observations such as the existence of a magnetic field further limit the possible makeup of the earth's interior.

Detailed studies of the propagation of seismic body waves have provided most of the information on the structure of the interior and the densities and physical states of the zones within it. Seismic data have made possible the identification of the silicate mantle, including a low-velocity layer underlain by several phase transitions in the upper mantle, and the much denser, iron-rich core. The S-wave shadow zone demonstrates that the outer core is liquid and indicates its size. High-pressure laboratory experiments and seismic data suggest that one or more elements besides iron and nickel are present at least in the outer core, reducing its density. Pressures in the interior are estimated from the densities of the various zones in the earth. Internal temperatures are constrained using the melting curves of ultramafic rocks (for the mantle) and of iron (for the core) for pressures corresponding to those prevailing in the interior.

Terms to Remember

cosmic abundance curve
phase changes
refraction
low-velocity layer
seismic shadow
Mohorovičić discontinuity (Moho)
seismic tomography
transition zone

Questions for Review

1. By what means do geoscientists obtain samples of the deep crust and uppermost mantle, below depths to which they can drill?
2. From what information is the cosmic abundance curve of the elements derived, and how is it used to constrain the composition of the earth's interior?
3. Can the earth's magnetic field be due to a lump of solid iron in the core? Why or why not? How else can the field be explained?
4. What is the Moho, and how is the depth to the Moho determined?
5. What causes the decrease in seismic velocities associated with the low-velocity layer?
6. Geoscientists do not believe that there are significant compositional changes in the lower mantle. Why not?
7. Describe the origin of the S-wave shadow zone. What does it tell geoscientists about the earth's interior?
8. How is the depth to the inner/outer core boundary determined?
9. The core is believed to be neither pure iron nor iron-nickel alloy. Why?
10. Can temperatures in the earth's interior be determined by extrapolation from geothermal gradients in the crust? How else are internal temperatures estimated? Discuss briefly.

For Further Thought

The relatively high cosmic abundance of iron suggests that iron is the dense element that forms the earth's core. In the absence of that constraint, consider whether one could demonstrate in some other way that the core is not made of another metal, such as gold or lead, and if so, how.

Suggestions for Further Reading

Bercovici, D., G. Schubert, and G. A. Glatzmaier. 1989. Three-dimensional spherical models of convection in the earth's mantle. *Science* 244:950–955.

Garland, G. D. 1979. *Introduction to geophysics.* 2d ed. Toronto: W.B. Saunders.

Gass, I. G., P. J. Smith, and R. C. L. Wilson, eds. 1971. *Understanding the earth.* Cambridge, Mass.: MIT Press. (See especially "Mohole: Geopolitical fiasco" by D. S. Greenberg, 342–48; "The composition of the earth" by P. Harris, 52–69; and "The earth's heat and internal temperatures" by J. H. Sass, 80–87.)

Henderson, P. 1982. *Inorganic geochemistry.* New York: Pergamon Press.

Hoffman, K. A. 1988. Ancient magnetic reversals: clues to the geodynamo. *Scientific American* 258 (May):76–83.

Lay, T., T. J. Ahrens, P. Olson, J. Smyth, and D. Loper. 1990. Studies of the Earth's deep interior: goals and trends. *Physics Today* 43 (October):44–52.

Scientific American 249 (September 1983). (Contains a series of articles on the earth and its various zones.)

Volcanoes and the earth's interior. 1983. San Francisco: W.H. Freeman. (A selection of readings from *Scientific American,* 1975–1982.)

York, D. 1975. *Planet earth.* New York: McGraw-Hill.

CHAPTER 12

Outline

The Continental Crust

Deformation of the continental crust has tilted these rocks at Split Mountain, Dinosaur National Monument, Utah.
© David C. Schultz

Introduction

The surfaces of the continents are the part of the earth most accessible for observation, measurement, and sample collection. In favorable cases, geologists can extrapolate features recognized at the surface some distance into the crust. Geologic structures are discussed in this chapter, not because they occur only in continental crust, but because they are best exposed, most readily examined, and most elaborately developed on the continents. This chapter describes common kinds of geologic structures and other features formed in the continental crust, and how they are identified; then briefly examines the nature of mountain building, and the different kinds of mountains; and concludes with a look at the continents through time, and what can be said about the longevity and growth of continents.

Geologic Structures

We have already looked at a variety of rock textures produced during the formation of a single rock unit—for example, the foliation of schists or the porphyritic or vesicular textures of some volcanic rocks. Additional structure can be imposed on rocks after formation, through the application of stress. The type of structure produced depends on the nature of the stress and on whether the rock behavior is more brittle or more plastic. As described previously, brittle behavior favors the formation of fractures; plastic behavior, folds.

Describing the Orientation of Structural Features

Frequently, what is important is not only the kind of geologic structure present, but also its orientation. This provides information about the directions from which the deforming stresses have come, which may, in turn, aid in tectonic interpretation. The orientation of lines and planes can be described in terms of two parameters: *strike* and *dip* (figure 12.1). **Strike** is a compass direction, measured parallel

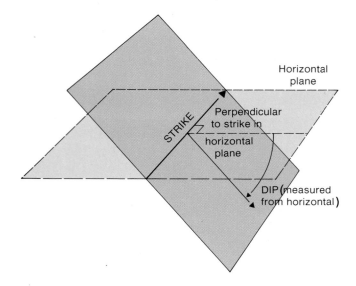

Figure 12.1   Strike and dip of a plane.

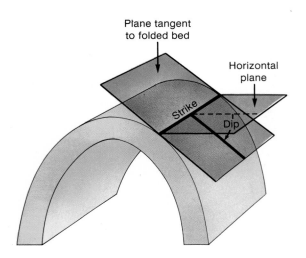

Figure 12.2   Strike and dip of a plane tangent to a folded bed.

to the earth's surface. **Dip,** as the name suggests, is displacement downward from the horizontal, measured in degrees. The strike of a plane is the compass orientation of the line of intersection of that plane with a horizontal plane. Dip is then measured perpendicular to the strike, by definition, and represents the maximum slope of the plane.

The concepts of strike and dip of a plane can be explored by immersing a sheet of stiff cardboard or a thin board partway into a basin of water. The strike of the board is the orientation of the line formed where board meets water surface; dip is measured at right angles to that direction. In an outcrop, the direction of dip of a sloping surface can be determined most readily by pouring

water on the outcrop: The water trickles down the steepest slope, which is also in the direction of dip. The strike is the trend of the horizontal line perpendicular to the dip.

Many geologic features are planar: sedimentary beds, igneous dikes and sills, metamorphic foliation. However, these planes may be deformed into nonplanar shapes that do not maintain the same orientation over broad areas. If a limestone bed is folded, for example, the bed will be curved, and its orientation becomes more complex to describe. Strike and dip can still be measured at any single point by considering the orientation of a hypothetical plane tangent to (just touching) the curved surface at that point (figure 12.2). To characterize the geometry of

Summary

Deformation of the continental crust takes the form of folds (plastic deformation) and faults (brittle deformation). The nature and orientation of these structures provide information on the orientation of the stresses that formed them. Both folds and thrust faults can result in horizontal crustal shortening and in crustal thickening.

The lower density and greater thickness of the continental crust relative to oceanic crust gives continents greater buoyancy and accounts for their relatively high relief. When a continental block is in isostatic equilibrium, the extra relief above sea level is compensated by a low-density root extending into the mantle. When erosion, orogenesis, glaciation, or other processes change the distribution of crustal mass, isostatic adjustment occurs, with the lithosphere rising or sinking as appropriate until equilibrium is restored.

Large mountain ranges are formed through orogenesis, which comprises deformation, metamorphism, and magmatic activity. The present topography of individual mountains is, in most cases, due to erosion by water or ice rather than to the nature of the crustal structures involved in their formation. Orogenesis most commonly occurs at convergent plate boundaries, where new material is accreted or sutured onto the margin of a craton. Some of this added material, now found in the so-called suspect terranes, has been transported over considerable distances. The fact that material is added to continental margins by successive orogenic events does not necessarily mean that there is a net increase in the total mass or extent of the continents through time, for continental material is also cycled back into the mantle through subduction.

Terms to Remember

accreted terrane
anticline
antiform
axial surface
axial trace
axis
craton
dip
dip-slip fault
fault
geosyncline
graben
hinge
horst
isostasy
isostatic
 equilibrium
joint

joint set
limbs
normal fault
oblique-slip fault
orogenesis
orogeny
recumbent fold
reverse fault
shield
strike
strike-slip fault
suspect terrane
suture
syncline
synform
terrane
thrust fault

Questions for Review

1. Define the strike and dip of a plane.
2. What is the distinction between joints and faults? Between a thrust fault and a normal fault?
3. Describe two processes by which continental crust may be thickened.
4. How are anticlines and synclines distinguished when they occur in sedimentary rock sequences?
5. Under what circumstances do overturned folds develop?
6. Explain the concept of isostatic adjustment, giving an example. Why is such adjustment not instantaneous?
7. Name and briefly describe three kinds of mountains.
8. Define the following terms: *craton, shield, orogeny.*
9. Summarize the principal orogenic processes at a continental margin with an adjacent subduction zone.
10. What is a *terrane?* A *suspect terrane?* An *accreted terrane?*

For Further Thought

1. The identification of suspect terranes is difficult, as demonstrated by their relatively recent discovery. Suppose that you have found a batholith adjacent to a sequence of interlayered sedimentary and volcanic rocks. What features or properties might you look for to determine whether or not these are two distinct terranes formed as different continental blocks that have become juxtaposed?
2. Would you expect, in general, that shield regions would be more or less likely to be in isostatic equilibrium than continental margins? Why?

Suggestions for Further Reading

Condie, K. 1989. *Plate tectonics and crustal evolution.* 3d ed. New York: Pergamon Press.

Davis, G. H. 1984. *Structural geology of rocks and regions.* New York: John Wiley and Sons.

Hobbs, B. E., W. D. Means, and P. F. Williams. 1976. *An outline of structural geology.* New York: John Wiley and Sons.

Hsu, K. J. 1982. *Mountain building processes.* New York: Academic Press.

Lyttleton, R. A. 1982. *The earth and its mountains.* New York: Wiley.

Meissner, R. 1986. *The continental crust: A geophysical approach.* Orlando, Fl.: Academic Press.

Miyashiro, A., K. Aki, and A. M. C. Sengor. 1982. *Orogeny.* New York: John Wiley and Sons.

Tarling, D. H. ed. 1978. *Evolution of the earth's crust.* New York: Academic Press.

Weyman, D. 1981. *Tectonic processes.* London: Allen and Unwin.

Windley, B. 1984. *The evolving continents.* 2d ed. New York: John Wiley and Sons.

Outline

The Ocean Basins

Spectacular development of coral in the Great
Barrier Reef of Australia results from a favorable
combination of factors, including warm tropical
waters and low input of sediment from rivers
draining the continent.

Introduction

For many centuries, the deep ocean basins were virtually unknown territory—dark, cold, inaccessible. That changed, however, with the development of echo-sounding and seismic methods for exploring the topography and structure of the sea floor. Later came sampling devices capable of collecting cores of sediment from the sea floor and dredging rocks off the bottom, and cameras to photograph the features of the deep. Still more recently, submersible research vessels, like miniature submarines, have allowed scientists to observe, sample, and photograph such features as ridge systems directly. This chapter briefly reviews the principal structural features of the ocean basins and the patterns of sediment distribution on the sea floor.

The Face of the Deep: Topographic Regions of the Sea Floor

If the oceans were drained of water, a complex topography would be revealed (figure 13.1). Far from being featureless, as they once were thought to be, the oceans contain mountains and valleys that rival in scale similar features on the continents. The sea floor can be subdivided into several main physiographic, or topographic, regions: the continental margins, the abyssal regions (broadly defined, the "deep ocean" where the bottom lies below 1,000 fathoms—6,000 feet, or nearly 2 kilometers), and the oceanic ridge systems. The characteristics of the continental marginal zones differ, depending on whether those margins are active or passive.

Abyssal Plains

The **abyssal plains** are essentially flat areas occupying what is also called the "deep ocean basin," away from continental margins and ridges. Except for the trenches, the abyssal plains constitute the deepest regions of the sea floor. Although the plains themselves are quite level, they are not featureless,

as is clear from figure 13.1. Often, they are dotted with **abyssal hills,** low hills rising several hundred meters above the surrounding plains, or the higher *seamounts*. These features, of volcanic origin, are described further later in the chapter, as they are not unique to the abyssal plains. Some abyssal plains are also cut by channels resembling shallow stream channels. The channels appear to be related to, or extensions of, the submarine canyons of many continental margins (also discussed later in the chapter). If so, they are probably carved by the same density currents, although it is difficult to understand how sea-bottom currents could continue to flow across nearly flat plains with sufficient velocity to cause very much erosion.

Passive Continental Margins

A **passive margin** is one that is seismically and volcanically quiet, not associated with an active plate boundary. A generalized profile across a passive continental margin shows that the marginal region consists of several zones distinguished, topographically, on the basis of slope (figure 13.2).

Immediately offshore from the continent is the **continental shelf,** a nearly flat, shallowly submerged feature. The average slope of the world's continental shelves is about one-tenth of a degree, or 10 feet of drop per mile of distance offshore. This accounts for the generally shallow waters above continental shelves, the majority of which lie no deeper than 100 fathoms (600 feet) at their outer limits. The average width of all continental shelves is about 70 kilometers (40 miles); the shelves at passive margins may extend nearly 1,000 kilometers (600 miles) offshore from the adjacent continent. The Atlantic and Gulf Coastal margins of North America are examples of passive margins.

The continental shelves are a natural site for the accumulation of sediment derived from the continent, and many shelves have quite flat surfaces as a consequence of the deposition of layer upon layer of sediment over the

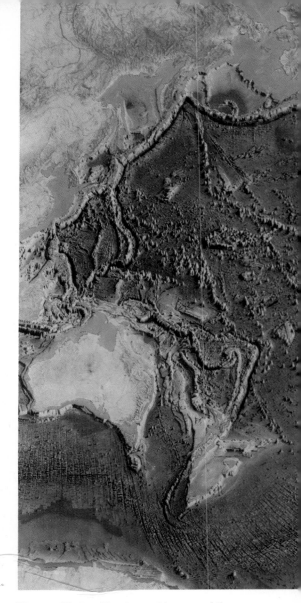

Figure 13.1 Physiographic map of the world ocean floor.
World Ocean Floor map by Bruce C. Heezen and Marie Tharp, 1977. Copyright © 1977 by Marie Tharp. Reproduced by permission of Marie Tharp, 1 Washington Ave., South Nyack, NY 10960.

underlying bedrock. This is not true of all shelves. During the last ice age, worldwide sea levels were lowered to such an extent that much of the now-submerged continental shelves was dry land. Rivers now terminating at the present shoreline flowed out tens or hundreds of kilometers across the continental shelves, carving valleys and depositing sediments just as they do now on the exposed continents. In regions directly subjected to glaciation, the ice sheets themselves carved up the shelf

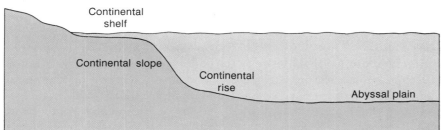

Figure 13.2 Cross section of the topography of a passive continental margin, showing continental shelf, slope, and rise.

and left behind deposits of sand, gravel, and boulders that now add topographic relief to the shelves. Martha's Vineyard and much of Cape Cod, for example, are glacial deposits.

The outer limit of the continental shelf is defined by the abrupt break in slope to the steeper **continental slope.** The angle of the continental slope varies widely, from nearly vertical to only a few degrees. The continental slope is the principal transition zone from near-continental elevations to deep-ocean depths. At the bottom (outside) edge of the continental slope, the topography again flattens out, the slope decreasing once more to less than 1 degree. The depth at which this occurs is highly variable, ranging from 1½ to 3 kilometers below sea level. The region beyond this second break in slope is the **continental rise.** The sea bottom continues to deepen across the rise until the deepest, flat *abyssal plains* are reached. (The continental rise in some areas is not well developed, and in these cases, the continental slope terminates in the deep ocean plains.)

Formation of Passive Margins

The nature and formation of passive continental margins can be understood in plate-tectonic terms. Passive

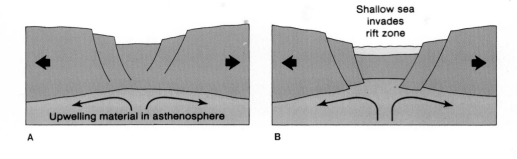

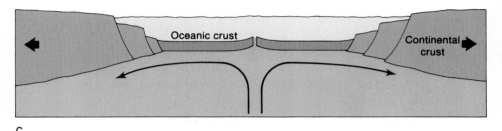

Figure 13.3 Formation of a passive margin. (*A*) The continent is stretched and fractured. (*B*) As rifting progresses, a shallow sea forms. Evaporation may leave salt deposits in the rift zone. (*C*) When rifting is complete, oceanic crust forms between the fragments of continent.

Figure 13.4 Satellite photograph of Lake Tanganyika, in the East African rift.
© NASA

margins are believed to have begun as continental rift systems, torn apart by mantle plumes and/or convection systems welling up from underneath (figure 13.3A). The continental lithosphere is stretched and thinned in the process. It ultimately fractures, along long faults perpendicular to the direction of stretching, and with continued tension, grabens form along the rift, making a long rift valley. This is now occurring in East Africa, which, in time, will split off from the rest of the continent, if rifting continues. Already, the floors of some of the graben-formed valleys of the East African rift system lie below sea level. They remain dry only because mountains presently block the ingress of the sea. Lake Tanganyika shows its rift origins in the complementary shapes of its east and west shores (figure 13.4).

Eventually, the rifting proceeds to the point that seawater floods the rift valleys (figure 13.3B). While the new seas remain shallow, rapid evaporation may cause deposition of salt deposits in these basins. This is the probable origin of the thick salt beds now buried along many continental margins.

With further tension, the rifting of the continent is completed (figure 13.3C). The rift valley is now floored by oceanic crust; an oceanic spreading ridge divides the fragments of continent; a narrow strip of true ocean basin lies between the separated continental masses. The Red Sea is in this stage of development. With continued spreading, the ocean widens, as the North Atlantic has following the breakup of Pangaea.

Throughout this process, the newly formed continental margins are passive in the sense that there is no subduction and therefore none of the associated seismic and volcanic activity. Limited volcanism does occur as magma works its way up through the fractured continent, and shallow earthquakes are associated with the downdropping of the grabens as well as with the spreading ridge when it has evolved. However, the active plate boundary in the system is the spreading ridge, once formed. Where the oceanic

and continental lithosphere meet at each margin is not a plate boundary; on a given side of the ridge, ocean and continent belong to the same plate. The passive margin is created by rifting, but once the intervening ocean is formed, further rifting is restricted to the sea-floor spreading ridge. Hence the subsequent tectonic quiescence of the continental margin so formed.

The lithospheric thinning and block-faulting together account for the deepening of the ocean seaward from the continent. The exact geometry of the faulting on a given margin exercises a fundamental control on the margin's topography. For example, the steep angles of continental slopes at some passive margins are probably fault scarps formed during the development of the margin. The topography is then modified by sedimentation. The extent to which sedimentary blankets soften the contours of the margin is principally a function of the quantity of sediment supplied from the continent. If the accumulated sedimentary sequence is a thick one, its weight may continue to depress the margin and deepen the sedimentary basin.

Active Continental Margins

Active margins are those characterized by ongoing seismic and/or volcanic activity; most are at or near subduction zones, and therefore, most are presently in the Pacific Ocean basin. Like passive margins, active margins have continental shelves, but the shelves adjacent to active margins are narrower than those at passive margins. The continental slope of an active margin is characteristically steeper than that of a passive margin, increasing in slope with depth. The slope of an active margin terminates not in a rise but in a **trench,** a long, deep, steep-walled valley trending approximately parallel to the continental margin. Figure 13.1 showed principal oceanic trenches. Trenches occur both at active continental margins and at ocean-ocean convergence zones, parallel to island arcs. The deepest trenches plunge to more than 10½ kilometers (6½ miles) below sea level. Trenches mark a plate boundary at which subduction is occurring and can serve as sites of sediment accumulation if there is adjacent land to serve as a sediment source.

Before the development of plate-tectonic theory, trenches were a major puzzle to marine geologists. No erosional mechanism to form them could be imagined. Seismic evidence and plate tectonics together suggested an explanation, when Benioff zones and subduction processes were recognized.

Certain specialized active margins arise in unusual tectonic settings. Subduction along the western edge of North America has caused a seafloor ridge system to be partially overridden by the continent. A portion of this margin, along the California coast, is now a transform fault, separating offset segments of spreading ridge off the coast of Oregon and Washington from other spreading segments in the Gulf of California.

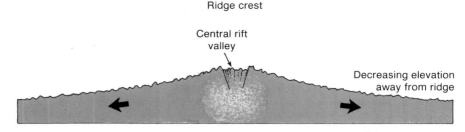

Figure 13.5 Topography of a spreading ridge in cross section (see also figure 13.1).

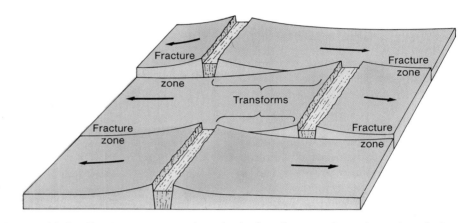

Figure 13.6 Fracture zones are the seismically quiet extensions of transform faults.

Oceanic Ridge Systems

Although the zones of active spreading are conventionally drawn as lines on a map, an oceanic ridge system is a broad feature (recall figure 13.1). The rift and its associated structures span wide areas of the sea floor. Figure 13.5 shows a topographic profile across an oceanic ridge system. In addition to the surface irregularities, two principal features may be observed: (1) the sea floor slopes away from the ridge crest symmetrically on either side, and (2) there is a valley at the very center of the ridge.

The valley is a **rift valley,** similar to a continental rift valley and formed in much the same way. The tensional stresses of rifting (spreading) cause deep, high-angle faulting, and with spreading, grabens bounded by these faults drop into the gap. The decreasing elevation of the sea floor away from the ridge was once thought to be due to relative uplift of the central portion of the ridge by upwelling mantle material. However, calculations have

shown that the profile can be explained by thermal contraction. Newly formed sea floor at the ridge, though solid, is warm. As it cools, it contracts. The slow cooling and contraction of the oceanic lithosphere as it moves away from the ridge at which it formed accounts for its declining elevation. Recall also from chapter 9 that the production and rifting of new sea floor at the ridge leads to symmetric patterns of magnetic anomalies across the ridge.

Chapter 9 already described the distribution of seismicity along a rift system, and the fact that shallow earthquakes are concentrated along the ridge proper and the transform faults between offset ridge segments. The fractures formed between offset segments do not disappear when plate movement has carried them beyond the spreading ridge (figure 13.6). They become seismically quiet where both plates move in the same direction, but the fractures remain, and there may be large scarps across the fractures. The fracture systems may be detectable outward from the rift over most of the region of elevated topography.

Along the ridge crest, seawater seeps into the young, hot lithosphere through fractures in the crust. The water is heated as it circulates through the warmed crust and reacts with the fresh seafloor rocks. It may escape again at **hydrothermal** (hot-water) **vents** along the ridge, rising up into the overlying colder seawater (figure 13.7). The escaping hot fluids are often clouded with fine suspended mineral particles; some have been nicknamed "black smoker chimneys" (figure 13.7B). The "chimneys," in turn, may be crusted with mineral deposits. Elements initially dissolved in the circulating waters are later deposited as the temperature and chemistry of the hydrothermal waters change upon reaction with fresh, cold seawater. Many of these minerals are metallic sulfides that could be important future mineral resources.

Other Features of the Ocean Basins

A variety of structural features—submarine canyons, hills of various sizes, and shallowly submerged platforms—occur in more than one of the major topographic regions of the ocean basins.

Submarine Canyons

A notable feature of many continental slopes at both active and passive margins is the presence of deep, steep-walled **submarine canyons** that cut across the shelf/slope system (figure 13.8). These canyons are usually V-shaped, sometimes winding; sometimes, they have tributaries on the continental shelves. Typically, they enlarge and deepen seaward, and extend to the base of the continental slope, where they often terminate in a fanlike deposit of sediment. They are cut into hard rock as well as into soft sediment.

The origin of submarine canyons has long been debated. A few are continuous with, and appear to be extensions of, major stream systems on land. Logically, it might be supposed that they

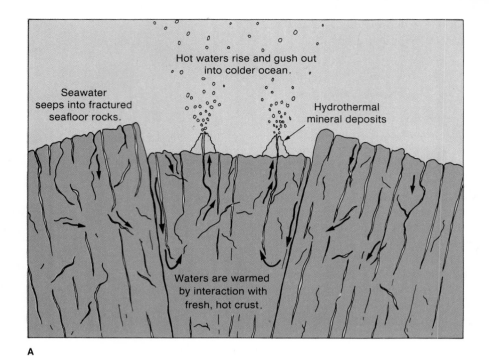

A

B

Figure 13.7 Seawater circulating through new seafloor at a spreading ridge produces hydrothermal vents. (A) Schematic diagram of circulation patterns. (B) A black smoker chimney.
(B) Photograph by W. R. Normark, courtesy of USGS Photo Library, Denver, Colorado

were cut by those stream systems at a time when the continental shelves were exposed by lowering of sea level. However, the canyons dissect not only the continental shelf, but in many cases, the whole width of the continental slope as well. Streams stop downcutting where they flow into the sea, and it seems highly unlikely that sea level would ever have been lowered to the base of the continental slope. Moreover, many submarine canyons have no corresponding stream systems on land; they seem to have formed independently of stream action. Another hypothesis attributed the canyons to submarine landslides, but this was not borne out by subsequent experiments and measurements; nor does it easily explain canyons in solid rock.

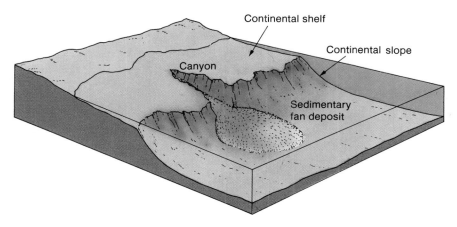

Figure 13.8 Topography of a submarine canyon.

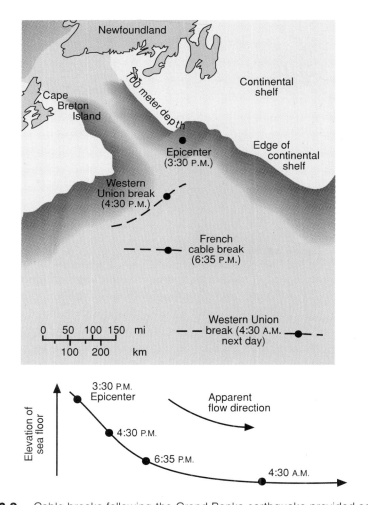

Figure 13.9 Cable breaks following the Grand Banks earthquake provided some of the strongest indirect evidence for turbidity currents.

From Shepard, Francis: *Earth Beneath the Sea.* The Johns Hopkins University Press, Baltimore/London, 1959, p. 25.

A possible mechanism of canyon carving was suggested, in part, by observations following an earthquake off the Grand Banks of Newfoundland in 1929. Several transcontinental telephone cables that had been run along the continental slope were broken. Furthermore, many of the breaks occurred hours after the earthquake and in apparent progressive sequence with increasing distance from the epicenter (figure 13.9). The overall time span in-

volved was too great for the breaks to be attributed directly to seismic-wave action. The most plausible explanation is that the breakage was caused by a fast-moving **turbidity current,** a density current of suspended sediment and water rushing down the continental slope. Turbidity currents are analogous to nuées ardentes: Just as the latter are denser-than-air ash clouds that flow along the ground as a consequence of their density, so the turbidity current is a denser-than-water mass that flows along the ocean bottom.

Fast-moving turbidity currents could be forceful enough not only to break underwater cables but also to carve canyons in rock or sediment. With the shallowing of slope at the continental rise, the currents would slow down and begin to deposit their sediment load. Individual beds of sediment at the bases of canyons often show graded bedding fining upward. This would be consistent with their deposition by gradual settling out of suspension, beginning with the coarsest material. Multiple graded beds, one atop another, could reflect successive turbidity flows.

A turbidity current could be formed as sediment is stirred up by an earthquake, or perhaps in conjunction with an undersea landslide. Since sediment is an essential component of turbidity currents, they are most common where there is an abundant sediment supply on the continental shelf. Turbidity currents may not account for all submarine canyons, but they appear to be the most plausible mechanism for forming the majority of the canyons. Indeed, those canyons not associated with continental river systems are difficult to account for in any other way.

Seamounts, Guyots, and Coral Reefs

The ridge systems and deep ocean basins are dotted with hills, some tall enough to break the surface and form islands, some not so high. These hills are generally of volcanic origin, whether associated with hot spots or with volcanism near the spreading ridges. The **seamounts** are those larger hills rising more than 500 fathoms

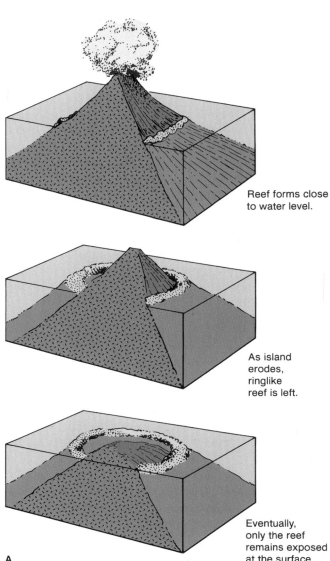

Reef forms close to water level.

As island erodes, ringlike reef is left.

Eventually, only the reef remains exposed at the surface.

A

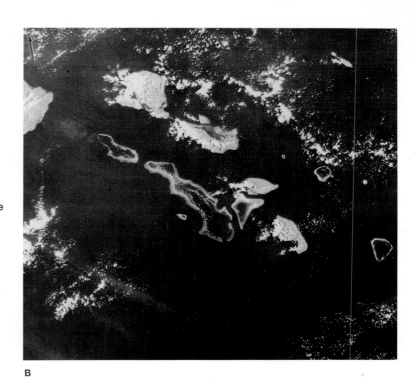

B

Figure 13.10 Coral atolls are closely related to marine volcanoes. (*A*) Schematic formation of an atoll. A reef forms around a volcanic island, which then erodes and/or sinks, leaving a circular atoll. (*B*) Satellite photograph of coral atolls in the Pacific. (*B*) © NASA

(3,000 feet, or about 1 kilometer) above the surrounding ocean floor, but not extending above the sea surface. Seamounts may or may not occur in linear chains (see again figure 13.1). Where a row of seamounts exists, age data may indicate formation in succession, as the oceanic plate moved over a stationary hot spot, as described in chapter 9.

A flat-topped seamount is called a **guyot.** The simplest explanation of the flat top is that a conical hill was planed off at sea level through wave action, but then how does the guyot come to be deeply submerged? Two possible mechanisms, operating separately or in conjunction, may account for the submergence of guyots. The first is related to the thermal contraction of sea floor

as it moves away from the ridge. As the sea floor cools and contracts, it sinks deeper and carries the guyot with it. A second mechanism is isostatic compensation: The mass of the guyot loads the lithosphere, which should sink or warp downward locally in response. Hot-spot volcanoes on the sea floor, becoming inactive as they move away from the hot spot, may evolve into guyots by this mechanism.

The formation of ringlike **coral atolls** is also linked to the existence of oceanic islands (figure 13.10). The corals require a solid foundation on which to build a reef. They also must be near the surface, in the lighted (*photic*) zone of the water, where the microscopic plants on which they feed,

and with which they coexist, are abundant. Reef construction then begins around the rim of an island. Weathering erodes the island. It may also be sinking, for reasons already described, or becoming submerged because sea level is rising worldwide. Reef growth continues upward, to keep the live colonies in the photic zone; the island's surface moves downward. Eventually, a large ring of reef may be all that projects to the water surface. Not all oceanic islands are hospitable to corals. Corals also require water warmer than 20° C (68° F), which further restricts their occurrence. (Fossil corals may therefore be sensitive paleoenvironmental indicators.)

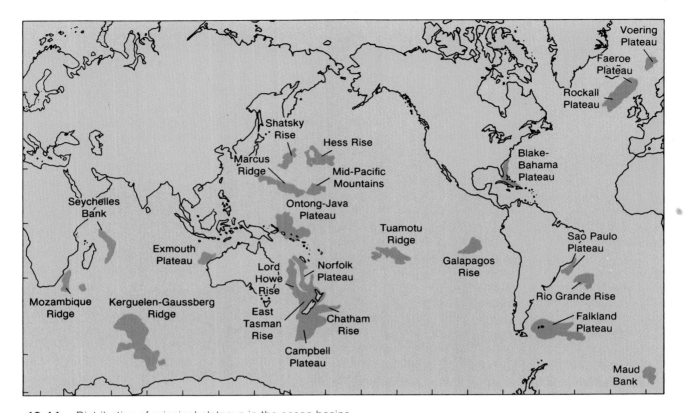

Figure 13.11 Distribution of principal plateaus in the ocean basins.

From Z. Ben-Avraham, A. Nur, D. Jones, and A. Cox, "Continental Accretion: From Oceanic Plateaus to Allochthonous Terranes," *Science* 213: 47–54. Copyright © 1981 by the AAAS. Reprinted by permission.

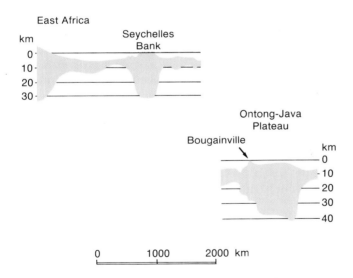

Figure 13.12 Some plateaus have a crustal structure that closely resembles that of continental crust.

Sources: Data from A. S. Furumoto, et al., "Crustal Structure of the Hawaiian Archipelago, Northern Melanesia, and the Central Pacific Basin by Seismic Refraction Methods," *Tectonophysics* 20: 153–164, 1973; and R. A. Scrutton, "Microcontinents and Their Significance," *Geodynamics Progress and Prospects*, ed. by C. L. Drake, pp. 177–189, American Geophysical Union, 1976.

Plateaus

In addition to ridge systems and sea-mount/island chains, the ocean basins contain broader **plateaus,** shallowly submerged or projecting partially above sea level. Some of the larger examples are mapped in figure 13.11; see also figure 13.1. A number of these plateaus consist of continental, not oceanic, crust, and seismic and other evidence shows that they are compensated isostatically by roots of continental crust, just as the continents themselves are (figure 13.12). The origin of these oceanic plateaus of continental affinity is unclear. The most likely explanation is that they are splinters off larger continents, perhaps split away by continental rifting during the formation of ocean basins.

Given the longevity of the sea floor (or lack of it) over geologic time, it seems that the plateaus will, sooner or later, be transported via plate movements to a subduction margin at which sea floor is being consumed. The

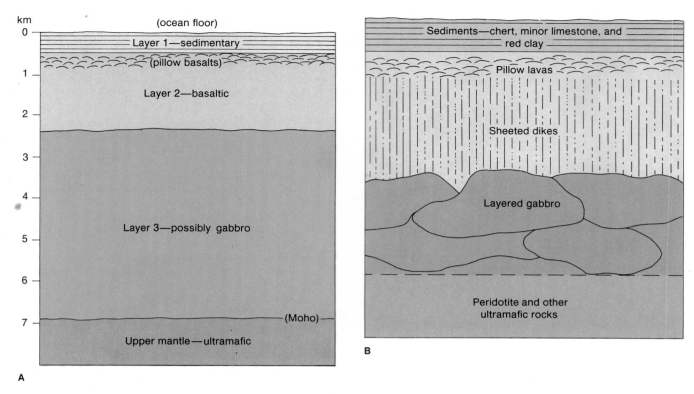

Figure 13.13 Ophiolites may be samples of the oceanic crust and upper mantle. (*A*) Typical cross section of the oceanic lithosphere. (*B*) Typical ophiolite sequence as reconstructed from outcrops on the continents.

greater buoyancy of continental lithosphere will prevent the plateaus' subduction. Instead, they might be pasted onto the edges of overriding continents. Perhaps the plateaus represent one possible origin of the suspect terranes found in the present continents.

The Structure of the Sea Floor; Ophiolites

Seismic evidence puts some constraints on the structure of the oceanic lithosphere (figure 13.13A). The oceanic crust is subdivided into three layers. Layer 1, which averages about 450 meters (1,500 feet) thick, consists of unconsolidated or unlithified sediment. This is known both from the low measured seismic velocities and from direct core sampling. Layer 2, an average of 1.75 kilometers (1.1 miles) thick, has also been sampled by deep coring and shown to consist of basalt. Layer 2 basalts often show a pillowed structure and glassy, quickly cooled rinds characteristic of submarine extrusion. This is entirely reasonable considering that Layer 2 would be at

the top of new oceanic lithosphere being created at a spreading ridge. The thickest layer of the oceanic crust, Layer 3, is about 4.7 kilometers (3 miles) thick. It has not been sampled. However, its seismic velocity is also consistent with a composition of basalt, or basalt's coarse-grained equivalent, gabbro. The base of Layer 3 is the Moho. Below that, seismic velocities jump to values typical of the ultramafic rocks of the mantle (probably peridotite, rich in olivine with some pyroxene).

Deep coring through the oceanic crust would be both technologically difficult and extremely expensive. However, samples of oceanic crust may be accessible on land, in the form of **ophiolites** (figure 13.13B). An ophiolite is a particular mafic/ultramafic rock association that is widely believed, on the basis of similarities between the types and arrangement of rocks in the ophiolites and analogous layers of the sea floor, to be a slice of old oceanic crust. Ophiolites contain pillow basalts like those of Layer 2. These are overlain by beds of *chert*, a microcrystalline

rock formed from silica-rich sediment; some areas of the sea floor today are sites of accumulation of siliceous sediment. Below an ophiolite's basalt flows is a complex of *sheeted dikes* (closely spaced, parallel, vertical dikes of basaltic composition) and gabbro. The gabbro would be the coarse plutonic rock crystallized at depth during the formation of new sea floor, with the sheeted dikes the feeder fissures of the basalt eruptions; abundant high-angle fractures would certainly be expected at a spreading ridge. Below the gabbros, ophiolites show ultramafic rock, plausible upper-mantle material. This ultramafic rock is typically greatly altered by weathering or low-grade metamorphism.

The layers in ophiolite sequences correspond closely with what is known of the sea floor, but the correspondence is not perfect. For example, the exact thicknesses of the individual layers in the ophiolite sequence are difficult to know, for ophiolites are commonly highly fractured, but the reconstructed thicknesses seem approximately right for oceanic crust.

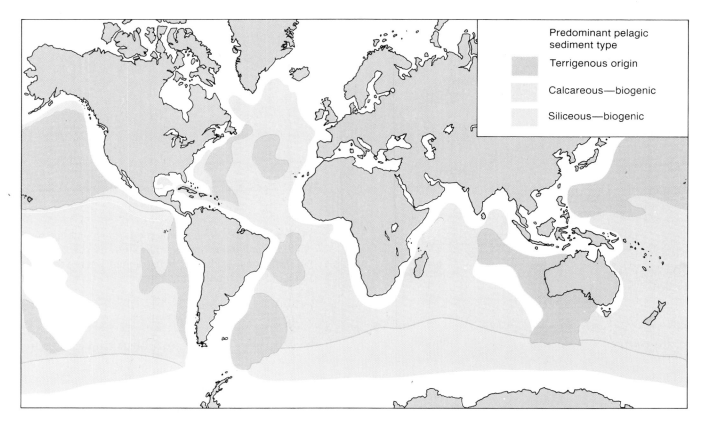

Figure 13.14 Distribution of principal sediment types on the sea floor.

After G. Arrhenius, "Pelagic Sediments," in *The Sea* by M. N. Hill, Vol. 3. Copyright © 1963 John Wiley & Sons, Inc., New York. Reprinted by permission of John Wiley & Sons, Inc.

Predominant pelagic sediment type

Terrigenous origin

Calcareous—biogenic

Siliceous—biogenic

Also, geologists cannot be sure that oceanic Layer 3 consists of sheeted dikes and gabbro; however, the seismic velocities are approximately right, and the sheeted dikes make sense tectonically. On balance, it seems highly likely that the ophiolites are indeed samples of oceanic crust and a bit of upper mantle.

It then remains to explain how denser oceanic crust comes to be perched on the continents. Ophiolites are exposed in mountain ranges. Mountains are commonly formed by convergence. It may be that, during convergence, especially when an ocean is slowly disappearing between two approaching continents, a few bits of ocean floor can be caught between converging plates and forced up onto a continent. This would be consistent both with where ophiolites are found and with their typically highly deformed condition. The process by which a slab of oceanic lithosphere is slipped up onto a continent has sometimes been termed **obduction,** to dis-

tinguish it from *sub*duction, in which the oceanic plate goes below the continent. On the basis of rock density, one would expect obduction to be relatively rare, and ophiolites are rather rare.

Sedimentation in Ocean Basins

Several sources supply sediment to the ocean basins. The principal sources are: (1) the continents, from which the products of weathering and erosion are transported to the sea by wind, water, or ice; (2) seawater, with its content of dissolved chemicals; and (3) the organisms that live in the sea, whose shells and skeletons provide a biogenic component to oceanic sediment. There is even a small but detectable contribution from the micrometeorites that rain unnoticed through the atmosphere. Figure 13.14 is a generalized map of the distribution of seafloor sediment types.

Terrigenous Sediments

Continental-marginal sediments are generally dominated by terrigenous material. **Terrigenous**—literally, "earth-derived"—**sediment** is clastic sediment derived from the continents. As already noted in chapter 6, the coarser sediments tend to be deposited first, closest to shore, with progressively finer materials transported farther seaward before they settle out of the water.

Sediments at the margins are not static, once deposited. They are redistributed by currents flowing along the continental shelf and slope, by turbidity currents and submarine landslides that carry them down the continental slope and beyond. The presence of a trench or rise near the margin may limit the extent to which terrigenous sediments, especially coarser ones, are transported away from the continent.

The chemical and mineralogical makeup of the terrigenous sediments is as varied as the weathering products

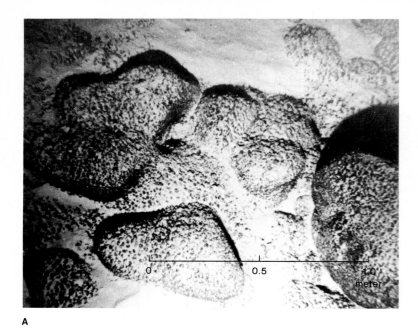

A

Figure 13.15 Where terrigenous sedimentation is limited, manganese nodules may litter the sea floor. (*A*) Manganese nodules, photographed where they have formed. (*B*) Distribution of manganese nodules in the ocean basins.

(*A*) Photograph by K. O. Emery, courtesy of USGS Photo Library, Denver Colorado. (*B*) Courtesy of G. Ross Heath and *Oceanus* magazine.

B

of the diverse rock types of the continents. The finest size fraction is typically dominated by clay minerals.

Pelagic Sediments

This finest size fraction stays in suspension most readily and may be transported beyond topographic barriers on the sea floor to settle into the abyssal plains. It is one component of **pelagic sediment.** Pelagic sediments (named from the Greek word *pelagos*, meaning "sea") are the sediments of the open ocean (away from continental margins), regardless of origin. As can be seen from figure 13.14, however, the pelagic sediments of much of the sea floor are dominated by material not of terrigenous origin.

Many of the larger marine creatures with shells or skeletons live in the shallower waters of near-shore regions. They contribute relatively little to pelagic sediment. Volumetrically, the most important biological contributors to pelagic sediment are microorganisms, the *foraminiferans* and *radiolarians.* The majority of species of foraminiferans and radiolarians have, respectively, calcareous ($CaCO_3$-rich) and siliceous (SiO_2-rich) hard parts. They live in the near-surface waters; when they die, they settle toward the bottom, the soft organic matter decaying away in the process. Although

each shell or skeleton is tiny, the immense numbers of organisms involved collectively produce a significant accumulation of sediment. The result is a fine-grained, water-saturated, calcareous or siliceous **ooze** on the bottom. The distribution of the biogenic oozes on the sea floor is only partially a reflection of the distribution of the corresponding organisms in near-surface waters. Both calcium carbonate and silica are also subject to dissolution in deep seawater, which selectively removes them from the sediments into solution.

> **R**ecall, for example, that calcite dissolves more readily in cold water than in warm. It also dissolves more readily at higher pressure. The ocean bottom waters are typically very cold and under high pressure from the overlying water column above. Calcite crystallized in warm shallow waters and then carried to the bottom begins to dissolve under the new pressure/temperature conditions.

Another locally important biogenic component of marine sedimentary rock is the calcareous reef, which may be built either by corals or by calcareous algae. As already noted, reefs must be built on a solid base, but within the photic zone. This, plus the effect of

temperature on calcium carbonate solubility, dictates that most reefs are found in warm, shallow waters, on continental shelves or shallowly submerged platforms. Corals also require relatively clear water to thrive; the flourishing ecosystem of Australia's Great Barrier Reef can be attributed, in part, to the fact that few stream systems drain into the ocean off the northeastern coast of Australia, so little terrigenous sediment is poured into those waters.

Other Oceanic Sediments

In relatively warm waters, such as are found on platforms, especially in tropical latitudes, carbonates may be precipitated directly out of seawater to form limestone beds. Some clays are precipitated directly from seawater also, although the bulk of marine clay is probably of terrigenous origin. And over much of the deep-sea floor, where overall sedimentation rates are slow, precipitation of manganese oxides and hydroxides forms the lumpy **manganese nodules** (figure 13.15). These may be an important mineral resource in the future, both for their manganese content and for the smaller amounts of other metals (cobalt, zinc, and nickel, for example) that they contain.

BOX 13.1

Ocean-Water Zonation, Upwelling, and El Niño

With the exception of turbidity currents, most of the vigorous circulation of the oceans is confined to the near-surface waters. Only the shallowest waters, within 100 to 200 meters of the surface, are well mixed, by waves, currents, and winds, and warmed and lighted by the sun. The average temperature of this layer is about 15° C (60° F).

Below the surface layer is the *thermocline,* a zone of rapidly decreasing temperature that extends 500 to 1,000 meters below the surface. Temperatures at the base of the thermocline are typically about 5° C (40° F). The thermocline is the interface between the warm, mixed layer and the so-called *deep layer* of cold, slow-moving, rather isolated water. The temperature of the bottommost water is close to freezing and may even be slightly below freezing (the water prevented from freezing solid by its dissolved salt content and high pressure). This cold, deep layer originates largely in the polar regions.

When winds blow parallel or nearly parallel to a coastline, the resultant currents may cause warm surface waters to be blown offshore. This, in turn, creates a region of low pressure and may result in **upwelling** of deep waters to replace the displaced surface waters (figure 1). The deeper waters are relatively enriched in dissolved nutrients, in part because few organisms live in the cold, dark depths to consume those nutrients. When the nutrient-laden waters rise into the warm, photic zone, they can support abundant plant life and, in turn, animal life that feeds on the plants. Many rich fishing grounds are located in zones of coastal upwelling.

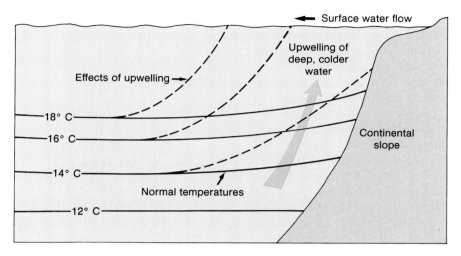

Figure 1 Upwelling results when strong offshore winds cause a low-pressure zone near shore and deep waters rise in response.

The west coasts of North and South America and of Africa are subject to especially frequent upwelling events. The productivity of fisheries off the coast of Peru is noteworthy.

From time to time, however, for reasons not precisely known, the upwelling is suppressed for a period of weeks or longer. Abatement of coastal winds, for one, would reduce the pressure gradient driving the upwelling.

The reduction in upwelling of the fertile cold waters has a catastrophic effect on the Peruvian anchoveta industry. Such an event is called *El Niño* ("the (Christ) Child") by the fishermen, for it commonly occurs in winter, near the Christmas season. The intensity of and interval between El Niño events, however, is quite variable. Significant El Niño conditions occurred in 1957–1958, 1965, 1972–1973, and 1982–1983. During each of these periods, various other meteorological problems arose worldwide—droughts in some places, torrential rains elsewhere. It may be that the same meteorological factors, including shifting wind patterns, cause both El Niño events and these abnormal weather conditions. This is a subject of very active current research.

Summary

The sea floor can be subdivided into continental-margin regions, spreading ridge systems, and abyssal plains. The continental margins are further divided, on the basis of tectonics, into active and passive margins. A passive continental margin, formed during continental rifting and creation of a new ocean, typically has a broad continental shelf; the angle of the continental slope is variable. An active, convergent margin has a narrow shelf, usually a steep continental slope, and a trench formed by subduction at the foot of the slope. The shelves and slopes are often dissected by canyons, most likely cut by turbidity currents. The abyssal plains are the deepest parts of the oceans. They may be dotted with hills, seamounts, and guyots. Oceanic ridge systems possess a central rift valley; the topography slopes away from the ridge crest, and fracture zones, extensions of transform faults, stretch for tens or hundreds of kilometers from the ridge crest. Hydrothermal activity occurs as a result of circulation of water through the hot rocks along the ridge.

The relatively thin oceanic crust is divided into Layer 1 (sedimentary), Layer 2 (basaltic), and an unsampled Layer 3 that may be gabbroic. Ophiolites may be slices of ancient, obducted oceanic crust. The crust underlying some of the large plateaus in the ocean basins, by contrast, has a density and thickness comparable to continental crust; these plateaus may be future accreted terranes. Terrigenous sedimentation in the oceans is largely confined to the continental margins. Pelagic sediment consists predominantly of fine clays (partially terrigenous) and calcareous and siliceous oozes. Possible future mineral resources from the sea floor include sulfides deposited at hydrothermal vents along the spreading ridges and manganese nodules of the deep sea floor.

Terms to Remember

abyssal hills
abyssal plains
active margin
continental rise
continental shelf
continental slope
coral atoll
guyot
hydrothermal
  vents
manganese
  nodules
obduction
ooze
ophiolite
passive margin
pelagic sediment
plateau (oceanic)
rift valley
seamount
submarine
  canyon
terrigenous
  sediment
trench
turbidity current
upwelling

Questions for Review

1. What are the abyssal plains? How flat and featureless are they?
2. What are the principal topographic features of a passive continental margin?
3. Describe the process of ocean-basin creation that leads to formation of a passive margin.
4. Where and how are trenches formed?
5. Sketch a topographic cross section of a spreading ridge, and account for the central valley and the general slope of the topography.
6. Describe the nature and probable origin of a turbidity current. What features of the sea floor might be explained by turbidity currents?
7. What is a guyot? Explain one mechanism by which guyots might be formed.
8. How do coral atolls develop?
9. The oceanic plateaus may be the suspect terranes of the future; explain briefly.
10. What is an ophiolite? Why are ophiolites believed to be samples of oceanic lithosphere, and how might they come to be found on the continents?
11. Cite and briefly explain two major components of pelagic sediment.
12. Name and describe two kinds of mineral deposits found on the sea floor that may become important in the future.

For Further Thought

1. In what ways might you be able to distinguish a sediment sample from the abyssal plain from a sediment sample collected at the base of the continental slope of a passive margin?
2. Suppose that, while drilling in the sediments on the continental shelf off New England, you find buried fossil coral reefs. The water there today is too cold for corals. Suggest at least two possible explanations for the presence of the fossil corals.

Suggestions for Further Reading

Bishop, J. M. 1984. *Applied oceanography.* New York: John Wiley and Sons.

Fanning, K. A., and F. T. Manheim, eds. 1982. *The dynamic environment of the ocean floor.* Lexington, Mass.: D.C. Heath.

Heezen, B. C., and C. D. Hollister. 1971. *The face of the deep.* New York: Oxford University Press.

Hill, M. N., ed. 1962–1983. *The sea.* New York: Interscience Publishers. (This eight-volume set, though somewhat dated in parts, still assembles in one place a wealth of data and basic information on the oceans and ocean basins.)

Kennett, J. P. 1982. *Marine geology.* Englewood Cliffs, N.J.: Prentice-Hall.

Pickard, G. L. 1979. *Descriptive physical oceanography.* 3d ed. New York: Pergamon.

Scrutton, R. A., ed. 1982. *Dynamics of passive margins.* Washington, D. C.: American Geophysical Union.

Scrutton, R. A., and M. Talwani, eds. 1982. *The ocean floor.* New York: Wiley-Interscience.

Seibold, E., and W. H. Berger. 1982. *The sea floor.* New York: Springer-Verlag.

Shepard, F. P. 1977. *Geological oceanography.* 3d ed. New York: Crane, Russak.

Walsh, J. J., ed. 1988. *On the nature of continental shelves.* New York: Academic Press.

Woods Hole Oceanographic Institution. 1984. El Niño. *Oceanus* 27 (no. 2, summer).

Woods Hole Oceanographic Institution. 1989. The oceans and global warming. *Oceanus* 32 (Summer).

Outline

This low-gradient, meandering stream, a tributary of the Grand River, flows slowly through the countryside of Prince Edward Island, Canada.
© David C. Schultz

Streams

Introduction

When rain falls on the earth's surface, or when accumulated snow melts, the water either sinks into the ground (**in-filtrates**) or remains at the surface. Subsurface waters are discussed in the next chapter.

Surface water tends to flow from higher to lower elevations. It may wash in broad sheets over the ground surface, but commonly, the flowing water is collected into a channel, forming a **stream.** The term *stream* is used to describe any body of flowing water confined within a channel, regardless of its size. In this chapter, we describe characteristics of streams and survey streams as agents of erosion, sediment transport, and deposition. Later in the chapter, we consider the phenomenon of stream flooding, and the interplay of flooding and human activities.

The Hydrologic Cycle

The **hydrosphere** includes all the water at and near the surface of the earth. Most of it is believed to have been out-gassed from the earth's interior early in its history, when earth's temperature was higher. Now, except for occasional minor contributions from volcanoes bringing up additional water from the mantle, the quantity of water in the hydrosphere remains essentially constant.

All of the water in the hydrosphere is caught up in the **hydrologic cycle,** illustrated in figure 14.1. The main processes in the cycle are evaporation into and precipitation out of the atmosphere. Precipitation onto land can re-evaporate (directly from the ground surface or indirectly through plants by transpiration), infiltrate into the ground, or run off over the ground surface. Both surface and subsurface runoff act to return water to streams and, usually, ultimately to the oceans. With their vast exposed surface area, the oceans are the principal source of evaporated water.

The total amount of water moving through the hydrologic cycle is large, more than 100 million billion gallons per year. A portion of this water is temporarily diverted for human use, but eventually, it makes its way back into the global water cycle by a variety of routes, including release of municipal sewage, evaporation from irrigated fields, or discharge of industrial wastewater into streams. Water in the hydrosphere may spend extended periods of time—even tens of thousands of years—in one or another of the water reservoirs, but from the longer perspective of geologic history, it is still regarded as moving continually through the hydrologic cycle.

Drainage Basins and Size of Streams

The geographic area from which a stream system draws water is its **drainage basin.** The volume of water in a stream at any spot is related, in part, to the size of the area drained, as well as to such other factors as the amount of precipitation in the area and how readily the water infiltrates into the ground. Typically, only very small drainage basins are drained by a single channel. As small streams flow into larger ones, drainage patterns become more complex.

One way to describe the complexity of drainage is in terms of stream *order* (figure 14.2). A **tributary stream** is one that flows into another stream. A *first-order stream* is one into which no tributaries flow; a *second-order stream* is one that has only first-order streams as tributaries; and so on. It is generally true that the higher-order the stream, the larger the area drained by the stream and its tributaries. The U.S. Geological Survey has identified between 1½ and 2 million streams in the United States. The more than 1½ million first-order streams have an average drainage area of 2½ square kilometers each; the 4,200 fifth-order streams average about 1,300 square kilometers in drainage area; the Mississippi River, the single tenth-order stream, drains 3.2 million square kilometers.

The majority of stream systems exhibit a drainage pattern described as **dendritic,** from the Greek word *dendros,* meaning "tree." In a dendritic drainage pattern, stream channels are irregular, and tributaries join larger streams at a variety of angles.

Local geology and the presence of regular structures may cause the drainage network to assume a more symmetric geometry. Where sets of parallel fractures cause zones of weakness and/or topographic lows, streams may establish a rectilinear **trellis drainage** pattern in which tributaries join the main stream at right angles (figure 14.3A). Trellis drainage may also develop where the topography is dominated by parallel ridges of resistant rock, and streams tend to flow between those ridges. Intersecting joint sets may create **rectangular drainage** with right-angle bends in all streams (figure 14.3B). Where there exists a conical topographic high, such as a volcano, **radial drainage** carries water away from the center of the high in all directions (figure 14.3C).

The size of a stream at any point can be defined in terms of its **discharge,** the volume of water flowing past a given point in a specified period of time. Discharge is calculated as the product of the cross-sectional area of the stream times the flow velocity (figure 14.4). Conventionally, discharge is measured in cubic feet per second or cubic meters per second. As a practical matter, discharge can be difficult to measure, for the velocity of the stream is not everywhere the same. Flow is generally slower along the channel bed and sides, where there is friction between water and channel, and faster near the center of the channel and surface of the stream. Discharge is a function of many factors, including climate (precipitation, availability of water), flow

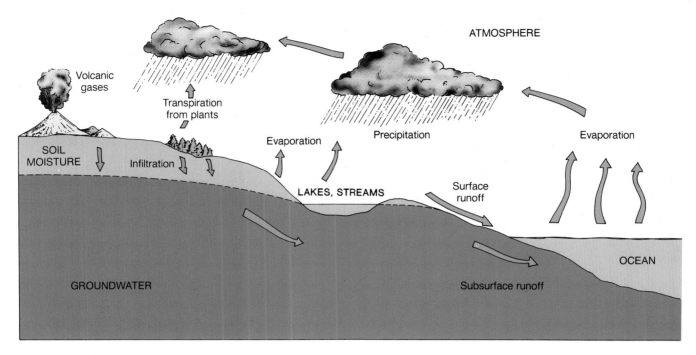

Figure 14.1   The hydrologic cycle.

Figure 14.2   Stream order is determined by the number of levels of tributaries flowing into the stream.

velocity, characteristics of the surrounding soil that influence the water infiltration, and the overall size of the drainage basin from which the streamflow is drawn. A stream's discharge, in turn, has a bearing on its effectiveness in erosion and sediment transport.

Streams As Agents of Erosion

Moving water is the principal erosive force at the earth's surface. Although it is most effective in eroding unconsolidated sediments, water can even scour solid rock. Stream erosion widens, deepens, and modifies the channel proper and, over time, alters the geometry of the valley surrounding the channel as well.

Streambed Erosion

A stream eroding downward into its bed is engaged in **downcutting.** How rapidly downcutting occurs is partly a function of the nature of the streambed material: Solid rock is generally much

A

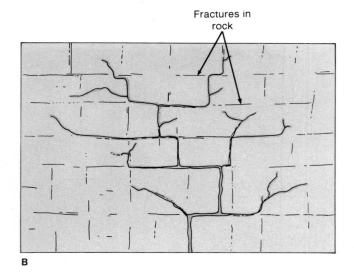

Fractures in
rock

B

C

D

Figure 14.3 Specialized geometry in drainage networks arises from special geologic conditions. (*A*) Fracture-controlled trellis drainage. (*B*) Rectangular drainage controlled by intersecting joint sets. (*C*) Radial drainage around a conical hill. (*D*) Dendritic drainage: no directional control due to structure or topography.

(*A*) M. R. Mudge, courtesy of USGS Photo Library, Denver, Colorado

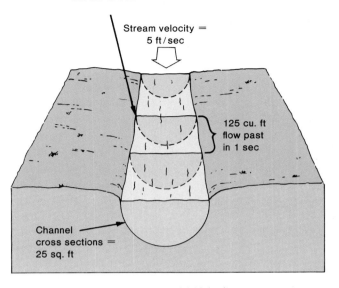

Discharge (flow past this point)
= (25 sq. ft.) × (5 ft/sec)
= 125 cu. ft/sec

Stream velocity =
5 ft/sec

125 cu. ft
flow past
in 1 sec

Channel
cross sections =
25 sq. ft

Figure 14.4 Discharge = Area × Velocity.

Figure 14.5 V-shaped valley of a rapidly downcutting stream: the Grand Canyon of the Yellowstone River.

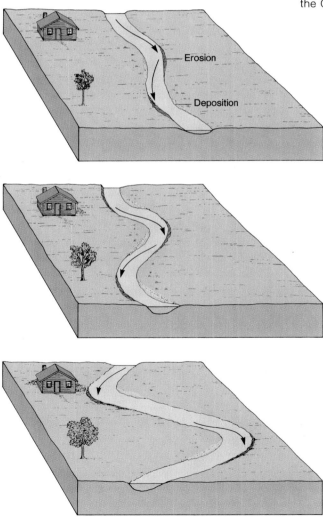

Erosion

Deposition

Figure 14.6 Map view of development of meanders. Channel erosion is greatest on the outside of curves, on the downstream side; deposition occurs in the sheltered area inside of curves. Through time, meanders migrate both laterally and downstream.

more resistant to erosion than is sediment. If the stream itself is transporting sediment, that sediment can abrade the channel bottom and sides, hastening erosion. The higher the velocity of stream flow, the more rapid the erosion, whether or not quantities of sediment are carried by the stream. The water can also attack soluble rocks chemically; a limestone, for example, is especially susceptible to dissolution. Where a stream is engaged in relatively rapid downcutting, it characteristically carves a narrow, steep-walled, V-shaped valley (figure 14.5).

Meanders

Where downcutting is less rapid and stream velocity slower, the straight channel begins to develop lateral displacements. Small irregularities in the channel cause local variations in flow velocity, which in turn contribute to variations in erosion and sediment deposition from point to point. Bends, or **meanders,** begin to develop along the channel (figure 14.6).

Once meanders begin to form, they tend to move both laterally and along the length of the stream. This can be understood in terms of flow velocity within the channel. The water strikes with greatest force the outer, downstream bank of the meander, which causes increased erosion in that area. Flow is slower on the inside and upstream bank of a meander, so sediment deposition tends to occur there, for reasons described later in the chapter. Over time, then, meanders can enlarge sideways and shift downstream by this combined erosion and sedimentation.

Meanders do not enlarge indefinitely. When large, looping meanders have developed along a stream, they represent sizeable detours for the flowing water. Eventually, the stream may make a shortcut, cutting off the meanders and abandoning the longer, irregular channel for a shorter, more direct route (figure 14.7). The cutoff meanders are termed **oxbows.** If they remain filled with water, they are *oxbow lakes.* Meander cutoff most commonly occurs during flood events, when the larger volume of water flowing faster in the channel may have enough additional momentum to carve the shortcut, bypassing the meander.

The Floodplain

Over time, the processes of lateral erosion associated with meandering, sediment deposition behind migrating meanders, and additional sedimentation during flood events when the stream has overflowed its banks collectively work to create a **floodplain** (figure 14.8). A floodplain is a flat or gently sloping region around a stream channel into which the stream flows during flood events (figure 14.9). At any given time, the meandering stream commonly occupies only a portion of the breadth of the floodplain. The overall width of the floodplain is a function of many factors, including the size of the stream, the relative rates of meander migration and of downcutting, and the strength of the valley walls.

While climatic and tectonic conditions remain fairly constant, slow lateral channel migrations over time tend to maintain a single floodplain with a fairly level surface. If these external conditions are suddenly changed significantly, relative rates of meandering

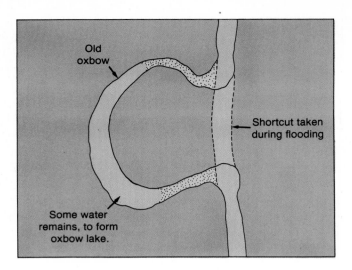

Figure 14.7   Map view of formation of oxbow by meander cutoff.

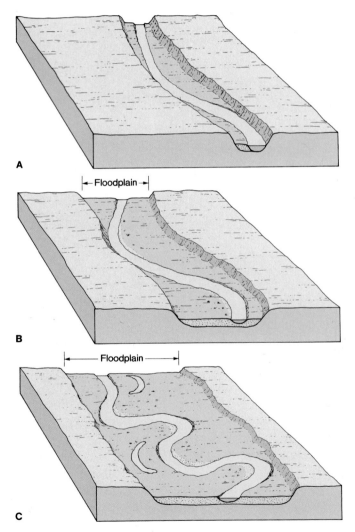

Figure 14.8   Meandering and sediment deposition during floods contribute to development of a floodplain. (*A*) Initially, the stream channel is relatively straight. (*B*) Small bends in the channel enlarge and migrate over time; meanders broaden. (*C*) Ultimately, a broad, flat floodplain is developed around the stream channel.

Figure 14.9 The meandering Sweetwater River, Wyoming, and its floodplain.

Photograph by W. R. Hansen, courtesy of USGS Photo Library, Denver, Colorado

Figure 14.10 Terraces within a valley, above the present floodplain.
© Arthur H. Doerr

Figure 14.11 Incised meanders in rock, with no floodplain surrounding the channel: goosenecks of the San Juan River in Utah.

Photograph by W. B. Hamilton, courtesy of USGS Photo Library, Denver, Colorado

and downcutting may also be altered in such a way as to change the consequent floodplain geometry. If a stream has established a broad floodplain and there is then regional uplift, for instance, downcutting may accelerate and the old floodplain be cut up by development of a new, narrower, deeper one within the broader valley of the old. This is one mechanism that has been proposed for the formation of **terraces,** steplike plateaus at some higher elevation above the present floodplain (figure 14.10). Multiple levels of terraces may be found within a single broad valley. Processes of terrace formation are complex and not fully understood.

Another somewhat problematic erosional process is the formation of **incised meanders** (figure 14.11). As their name suggests, these are meanders deeply cut into rock. The stream is not surrounded by a floodplain at nearly the same elevation but is entrenched within a deep, winding, steep-sided valley closely matching the meandering course of the stream. Several modes of formation have been proposed, and it is not always possible to determine the one responsible in each instance. Some streams, especially slowly flowing ones on gently sloping surfaces, exhibit meanders from their inception, and if downcutting is relatively rapid and the valley wallrocks strong enough to maintain steep cliffs, these early meanders may be dug deeper and deeper. Alternatively, meanders may be developed initially in soft or unconsolidated materials and subsequently fixed in form as the stream cuts down below the softer rock into rocks less prone to lateral erosion. The meander pattern in this instance is imposed on the underlying rock after formation in softer layers above.

Sediment Transport and Deposition

Streams can move material in several ways. The heaviest debris may be rolled or pushed along the streambed. This material is described as the **bed load** of the stream. The stream's **competence** at any point is the largest size of particle it can move in the bed load, which

Figure 14.12 Suspended sediment clouds the tributary at right. Note that the two streams do not immediately mix where they join. Yellowstone National Park.

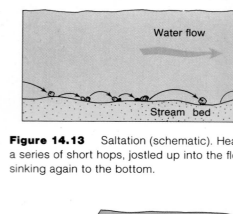

Figure 14.13 Saltation (schematic). Heavier particles move in a series of short hops, jostled up into the flowing water, then sinking again to the bottom.

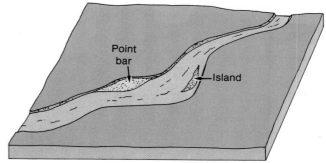

Figure 14.14 Point bars and islands form in a sediment-laden stream where the water flow is slowed.

is largely a function of its velocity. Stream **capacity** is the total quantity of material that a stream can potentially move, which depends on discharge (that is, on both velocity and quantity of water present). The **suspended load** consists of material that is light or fine enough to be moved along suspended in the stream, supported by the flowing water. The maximum suspended load that can be transported is also dependent on discharge. Suspended sediment clouds a stream and gives the water a muddy appearance (figure 14.12). Material of intermediate size may be carried in short hops along the streambed by a process called **saltation** (figure 14.13). Finally, some substances can be completely dissolved in the water (**dissolved load**).

The total quantity of material that a stream transports by all these methods is called, simply, its **load.** How much of a load is actually transported depends on the availability of sediments or soluble material: A stream flowing over solid, insoluble bedrock is not able to dislodge much material, while a similar stream flowing through sand may move a considerable load.

When the stream eventually deposits its load, the resulting deposit is called **alluvium,** a term derived from the French and Latin for "to wash over." Alluvium is a general term for any stream-deposited sediment. Certain alluvial deposits take distinctive forms that are given more specific names.

Velocity and Sediment Sorting

Variations in a stream's velocity along its length are reflected in the sediments deposited at different points. The sediments found motionless in a streambed at any point are those too big or heavy for that stream to move at that point. Where the stream flows most quickly, it carries gravel and even boulders along with the finer sediments. If velocity controls the maximum particle size that can be moved, it follows that, as the stream slows down, it selectively drops the heaviest, largest particles first, continuing to move along the lighter, finer materials. As velocity continues to decrease, successively smaller particles are dropped. In a very slowly flowing stream, only the finest sedi-

ments and dissolved materials are still being carried. This link between the velocity of water flow and the size of particles moved accounts for one common characteristic of alluvial sediments in channels: They are often well sorted by size or density, with materials deposited at a given point tending to be similar in size or weight.

Depositional Features

As a meander migrates laterally, eroding its outer bank, there generally is corresponding deposition along the inner bank, where the water flow is slower. The sedimentary feature thus built up is a **point bar** (figure 14.14). Where water slows along the channel bed, through friction, it may blanket the bottom with sediment. If the slowdown is a localized one—caused, for example, by obstacles such as isolated boulders on the bottom—the sediment deposition may be correspondingly localized into an island within the channel. Once such a feature has formed, it further impedes stream flow and may therefore tend to enlarge as water continues to be slowed and more

Figure 14.15 Braided streams, with many channels dividing and rejoining. This is typical of sediment-choked streams in glacial outwash, as here in Denali National Park, Alaska.
© NASA

sediments are deposited. Bars are dynamic features that can migrate or can be washed out completely by the high discharge of the stream's next flood.

Where the sediment load carried is large in relation to capacity, channel islands may grow until they reach the water surface, where they may be further stabilized by vegetation. The stream channel then has become divided into two channels, a process called *braiding*. A **braided stream** is one with several (perhaps many) channels that divide, then rejoin downstream (figure 14.15). Patterns of braided channels may become very complex. Where many shallow, braided channels cross a broad area of easily eroded sediment, the channels may shift constantly and the patterns change daily or even hourly.

A floodplain is itself a depositional feature, in part. As channels shift, point bars build land where stream waters flowed. Additional deposition in the floodplain can occur outside the channel, during flood events. The whole floodplain can be blanketed in sediment as the channel overflows and the flood waters are slowed by vegetation or other obstacles. Such sediments are descriptively termed **overbank deposits.** Close to the channel, the over-

bank deposits may assume a distinctive form, making ridges along the edge of the channel. These are natural **levees.** Development of levees is generally most pronounced where the suspended sediment load is relatively coarse. Streams carrying primarily fine, silty sediments may not form obvious levees.

As previously noted, a decrease in stream velocity is typically accompanied by deposition, beginning with the coarsest sediments. An abrupt drop in velocity may result in a large deposit. This is not uncommon at the **mouth,** or end, of a stream, where the stream flows into another body of water or terminates in a dry valley or plain.

Sedimentary deposits at the mouths of streams are characteristically fan- or wedge-shaped. Such a deposit, formed where a tributary stream flows into another, more slowly moving stream, is an **alluvial fan.** Alluvial fans also form where fast-moving mountain streams flow out onto plains or into flatter valleys, with a consequent sharp drop in water flow velocity (figure 14.16).

If a stream flows into a body of standing water, such as a lake or ocean, and the flow velocity drops to zero, all but the dissolved portion of the load is deposited, and the resulting sediment wedge is called a **delta** (figure 14.17).

Over time, deposition of delta sediments may actually build the land outward into the water.

Stream Channels and Equilibrium

For stream waters to flow, there must be a difference in elevation between the **source,** where the first perceptible channel indicates the stream's existence, and the mouth. The **gradient** of the stream is a measure of the steepness of the channel's slope. It is the difference in elevation between two points along the stream divided by the length of the channel between them. The higher the gradient, the steeper the channel and, all else being equal, the higher the water velocity.

The water level at the mouth of a stream, where it generally flows into an ocean, lake, or other stream, is the stream's **base level.** Base level represents the lowest level to which a stream can cut down. Once elevation upstream reaches base level, the gradient is zero, velocity should become zero, and no further erosion is possible.

Streams are dynamic systems, their forms changing in response to changes in environmental factors. If, over time, regional precipitation and stream discharge increase, increased erosion will scour a larger channel to accommodate the larger volume of water. An influx of sediment may result in increased sedimentation. If land in the drainage basin is uplifted, the increasing stream gradient and the consequent increases in velocity and rate of downcutting will hasten the stream's reapproach to its base level.

Equilibrium Profile

Over time, provided that environmental factors remain approximately constant, streams and their channels tend to adjust to the prevailing conditions and approach **dynamic equilibrium.** In terms of erosion and deposition, this means that sediment transport is just sufficient to balance the input of sediment from the drainage basin, and erosion of and deposition in the stream channel are

Figure 14.16 Alluvial fan deposited where fast-moving, sediment-bearing mountain stream flows into a plain and gradient decreases sharply. Rocky Mountain National Park.

Figure 14.17 The Mississippi River delta (satellite photograph). Suspended sediment clouds the water beyond.
© NASA

equal. A stream in which the channel and its gradient have been thus adjusted, so that input and outflow of sediment are just equal, is sometimes described as a **graded stream.**

A graded stream tends to show a characteristic pattern of decreasing gradient from source to mouth, illustrated in the **longitudinal profile** of figure 14.18. Generally, the gradient is steepest near the source, flattening toward the mouth.

In natural streams, velocity does not necessarily decrease downstream, despite the decreasing gradient. The explanation lies in the increase in water volume downstream, associated with the addition of tributaries to the stream and increasing drainage area. The added weight of water being pulled down toward base level by gravity partially compensates for the decrease in gradient. Overall discharge does increase downstream.

The geology of natural systems can cause deviations from this idealized profile. A common deviation is the occurrence of **knickpoints,** sudden drops in elevation of the channel bed. They can arise from stream flow across different rock types of different erodability, or by faulting; also, above the junction with a tributary, the channel may be cut less deeply than below, where an abruptly greater volume of water is present. Major knickpoints may be marked by waterfalls (figure 14.19). Once established, such scarps tend to erode headward, or upstream, as continued bed erosion wears them back or as material is carried downstream from the base of the scarp and the undercut scarp slumps and fails. The upstream migration of Niagara Falls is a classic example of such **headward erosion** of a knickpoint.

The concept of a stream in perfect equilibrium is an abstraction. Over the long term, gravity tends to level the land, and ultimately, in the absence of tectonic uplift, volcanic eruption, or other mountain-building activities, a net downcutting should be expected until the land is flattened to sea level.

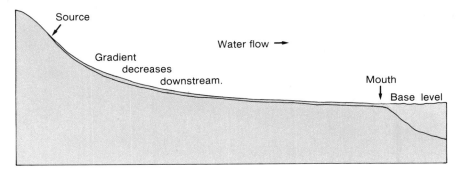

Figure 14.18   Longitudinal profile of a graded stream, concave upward, with steeper gradient near the source.

Figure 14.19   Niagara Falls exists because the erosion-resistant Lockport Dolomite has created a knickpoint.

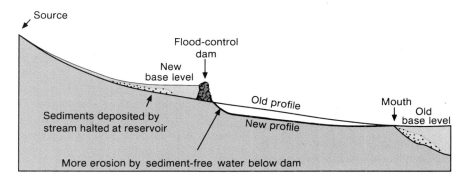

Figure 14.20   Effects of a dam and reservoir on a stream profile. The new base level of the reservoir causes deposition in the reservoir and upstream from it. Erosion increases below the dam.

However, the approach to equilibrium can be so close that, on a human time scale, the net change is negligible.

Effects of Changes in Base Level; Dams and Reservoirs

Tectonic uplift can raise the upper portion of a drainage basin relative to the base level of the stream system (an adjacent ocean, for example). The net effect is to increase the overall gradient of the system and the rate of downcutting toward the base level. Also, changes in sea level, such as are caused by glaciation (see chapter 17), move base level up or down and decrease or increase average gradients accordingly.

When human activities alter base levels, other modifications of the stream system are possible. For example, if a dam with a reservoir behind it is built along a previously graded stream, the water level in the reservoir becomes a new local base level for the stream above the dam. The stream's average gradient is decreased, and any downcutting should likewise decrease. Moreover, when the stream reaches the still reservoir, and flow ceases, the stream drops its (nondissolved) sediment load. This begins to fill in the reservoir, reducing the volume available for water storage, whether for irrigation, water supply, flood control, or whatever. Below the dam, the water released is free of suspended sediment or bed load, and is therefore capable of increased erosion through uptake of sediment. The net result of all of these changes is reflected in a modified longitudinal profile, as shown in figure 14.20.

Flooding

In most (moderately humid) climates, the size of a stream channel adjusts to accommodate the average maximum annual discharge. Much of the year, the water level (stream **stage**) may be well below the stream bank height. From time to time, the stream will reach **bankfull stage.** Should the water rise above bankfull stage, overflowing the channel, the stream is at **flood stage.**

Floods occur, in other words, when the rate at which water reaches some point in a stream exceeds the bankfull discharge there. Floods are not unnatural or particularly unusual events. Instead, flooding is a perfectly normal and, to some extent, predictable phenomenon.

Factors Influencing Flood Severity

Many factors together determine whether a flood will occur. The quantity of water involved and the rate at which it is put into the stream system are among the major factors. In the United States, heavy rainfall events occur in the southeastern states, which are vulnerable to storms from the Gulf of Mexico; the western coastal states, which are subject to prolonged storms from the Pacific Ocean; and the mid-continent states, where hot, moist air from the Gulf of Mexico collides with cold air sweeping down from Canada. Streams that drain the Rocky Mountains are particularly likely to flood during snowmelt events, especially when rapid spring thawing follows a winter of unusually heavy snow.

The risk of flooding may be moderated by infiltration, the rate of which, in turn, is controlled by the soil type and how much soil is exposed. Soils, like rocks, vary both in proportion of pore space and in the rate at which water seeps through them (*permeability*). If the soil is less permeable, the proportion of water that runs off over the surface increases. Also, once even permeable soil is saturated with water and all the pore space is filled (as, for instance, by previous storms), any additional moisture is necessarily forced to become part of the surface runoff. Similarly, an unexpected rainstorm in midwinter that dumps rain onto solidly frozen ground is more likely to cause flooding than is a similar storm in warm weather, when the ground is thawed. Topography, too, influences the extent of surface runoff: The steeper the terrain, the more readily water runs off over the surface and the less it tends to sink into the soil.

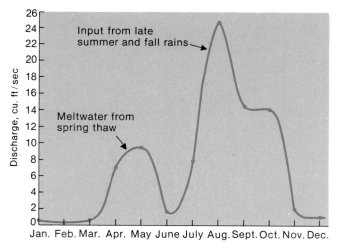

Figure 14.21 Sample hydrograph for Horse Creek near Sugar City, Colorado, spanning a one-year period.
Source: *U.S. Geological Survey Open-File Report 79–681.*

Water that infiltrates into the soil tends to flow downhill, like surface runoff, and may, in time, also reach the stream. However, the underground runoff water, flowing through soil or slowly than the surface runoff. The more gradually the water reaches the stream, the better the chances that the stream discharge will be adequate to carry the water away without flooding. Therefore, the relative amounts of surface and subsurface runoff, which are strongly influenced by the near-surface geology of the drainage basin, are fundamental factors affecting the severity of stream flooding.

Vegetation may reduce flood hazards in several ways: by providing a physical barrier to surface runoff, slowing it down; through plants' root action, which keeps the soil looser and more permeable, thereby increasing infiltration and decreasing surface runoff; and by soaking up some of the water and later releasing it by evapotranspiration from foliage. Vegetation can also be critical to preventing soil erosion. When the vegetation is removed and erosion increased, much more soil may be washed into streams. There, it can begin to fill in the channel, decreasing channel volume and thus reducing the stream's ability to carry water away quickly.

Parameters for Describing Flooding

During a flood, stream stage, velocity, and discharge all increase as a greater mass of water is pulled downstream by gravity. The magnitude of the flood can be described either by the maximum discharge measured or by the maximum stage. The stream is said to **crest** when the maximum stage is reached. This may occur within minutes of the influx of water, as with flooding just below a failed dam. However, in places far downstream from the water input, or when surface runoff has been delayed, the stream may not crest until several days after the flooding episode begins. In other words, the worst is not necessarily over just because the rain has stopped.

Fluctuations in stream stage or discharge through time can be plotted on a **hydrograph** (figure 14.21). Hydrographs spanning long periods of time are very useful in constructing a picture of the normal behavior of a stream and of that stream's response to flood-causing events. A flood shows as a peak on the hydrograph. The height and width of that peak and its position in time relative to the water-input event(s) depend, in part, on where the measurements are being taken, relative to where the excess water is entering the

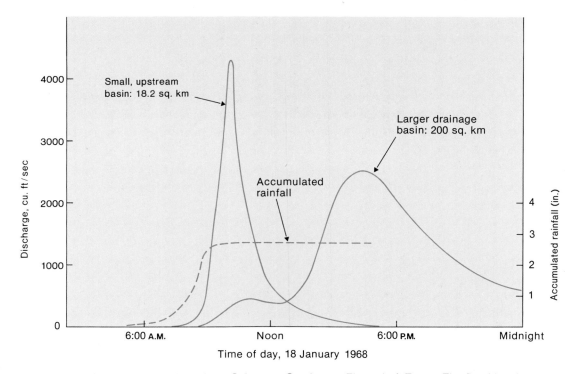

Figure 14.22 Flood hydrographs for two points along Calaveras Creek near Elmendorf, Texas. The flood has been caused by heavy rainfall. Upstream, flooding quickly follows rain. Downstream in the basin, response to water input is more sluggish.
Source: U.S. Geological Survey Water Resources Division.

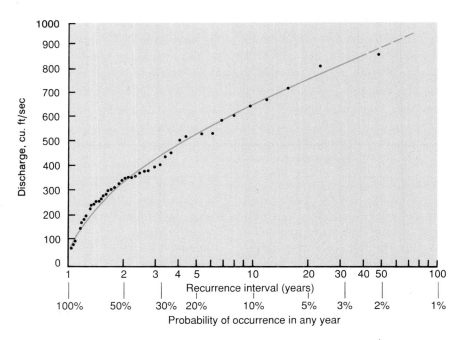

Figure 14.23 Sample flood-frequency curve for the Eagle River at Red Cliff, Colorado. Records span forty-six years, so assessments of likelihood of moderate flood events are probably fairly accurate.
Source: *U.S. Geological Survey Open-File Report 79–1060.*

system (figure 14.22). Measurements made near the point where the excess water is entering the system will show an earlier, sharper peak. By the time that water pulse has moved downstream to a lower point in the drainage basin, it will have dispersed somewhat, so that the peak on the hydrograph will be later, lower, and of longer duration. A short event like a severe cloudburst tends to produce a sharper peak than a more prolonged event like several days of steady rain or snowmelt, even if the same amount of water is involved.

Flood-Frequency Curves

Another way of looking at flooding is in terms of the frequency or probability of flood events of differing severity. The availability of long-term records makes it possible to construct a curve showing discharge as a function of recurrence interval, or probability of occurrence, for a particular stream or section of one (figure 14.23). A given flood event can then be described by its **recurrence interval** (how

BOX 14.1

How Big Is the 100-Year Flood?

The usual way of estimating the recurrence interval of a flood of given size is as follows: Suppose that records of maximum discharge (or stage) reached by a particular stream each year have been kept for N years. Each of these yearly maxima can be given a rank M, ranging from 1 to N, 1 being the largest, N the smallest. Then the recurrence interval R of a given annual maximum is defined as:

$$R = (N + 1)/M$$

For example, table 1 shows the maximum one-day mean discharges of the Big Thompson River, as measured near Estes Park, Colorado, for twenty-five consecutive years, 1951–1975. If these values are ranked, 1 to 25, the 1971 maximum of 1,030 cubic feet/second is the seventh largest and, therefore, has an estimated recurrence interval of $(25 + 1)/7 = 3.71$ years.

Suppose, however, that only ten years of records are available, for 1966–1975. The 1971 maximum discharge happens to be the largest in that period of record. On the basis of the shorter record, its estimated recurrence interval is $(10 + 1)/1 = 11$ years. Alternatively, if we look at only the first ten years of record, 1951–1960, the recurrence interval for the 1958 maximum discharge of 1,040 cubic feet/second (a discharge nearly the same size as the 1971 maximum) can be estimated at 2.2 years. Which estimate is right?

Table 1 Calculated recurrence intervals for discharges of Big Thompson River at Estes Park, Colorado

Year	Maximum mean one-day discharge (cu. ft/sec)	For twenty-five-year record M (rank)	R (years)	For ten-year record M (rank)	R (years)
1951	1,220	4	6.50	3	3.67
1952	1,310	3	8.67	2	5.50
1953	1,150	5	5.20	4	2.75
1954	346	25	1.04	10	1.10
1955	470	23	1.13	9	1.22
1956	830	13	2.00	6	1.83
1957	1,440	2	13.0	1	11.0
1958	1,040	6	4.33	5	2.20
1959	816	14	1.86	7	1.57
1960	769	17	1.53	8	1.38
1961	836	12	2.17		
1962	709	19	1.37		
1963	692	21	1.23		
1964	481	22	1.18		
1965	1,520	1	26.0		
1966	368	24	1.08	10	1.10
1967	698	20	1.30	9	1.22
1968	764	18	1.44	8	1.38
1969	878	10	2.60	4	2.75
1970	950	9	2.89	3	3.67
1971	1,030	7	3.71	1	11.0
1972	857	11	2.36	5	2.20
1973	1,020	8	3.25	2	5.50
1974	796	15	1.73	6	1.83
1975	793	16	1.62	7	1.57

Source: Data from U.S. Geological Survey Open-File Report 79–681.

frequently a flood of that severity occurs, on average, for that stream). For the stream of figure 14.23, for example, a flood with discharge of 675 cubic feet/second is called a *10-year flood*, meaning that a flood of that size occurs about once every ten years. An alternative way of looking at flooding is in terms of probability, which is inversely related to recurrence interval. That is, a 10-year flood has a 1/10 (or 10 percent) probability of occurrence in any one year.

Flood-frequency curves can be extremely useful in assessing regional flood hazards. In principle, even if a 100- or 200-year flood event has not occurred within memory, planners could, with the aid of topographic maps, project how much of a region would be flooded in such events. This, in turn, would be useful in siting new construction projects so as to minimize the risk of flood damage, or in making property owners in threatened areas aware of dangers. Unfortunately, in the absence of long-term records, the accuracy of the flood-frequency curve may be low (see box 14.1).

Another complication is that streams in heavily populated areas are being affected by human activities, many of which aggravate flood hazards. The flood-frequency curves are therefore changing with time. What may have been a 100-year flood two centuries ago might be a 50-year flood today.

It is important to remember that the recurrence intervals assigned to floods are *averages*. Over many centuries, a 50-year flood should occur an average of once every fifty years. However, that does not mean that two 50-year floods could not occur in successive years, or even in the same year. Therefore, it would be foolish to assume that, just because a severe flood has recently occurred in an area, the area is in any sense "safe" for awhile.

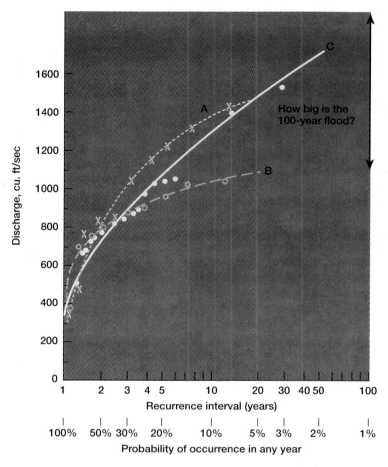

It is rare to have one hundred years or more of records for a given stream, so the magnitude of 50-year, 100-year, or larger floods is commonly estimated from a flood-frequency curve. Curves A and B in figure 1 are based, respectively, on the first and last ten years of data for the Big Thompson River (see last two columns of table 1). These two data sets give estimates for the size of the 100-year flood that differ by more than 50 percent. These results, in turn, both differ from an estimate based on the full twenty-five years of data (curve C). Figure 1 graphically illustrates how the availability of long-term records in the projection of recurrence intervals of larger flood events can smooth out short-term anomalies in streamflow patterns.

Figure 1 Short-term records (curves A and B) can be misleading about true recurrence intervals or probabilities of floods of different magnitudes. Compare with curve C, based on a twenty-five-year record.

Flooding and Human Activities

Urbanization can increase flood hazards in a variety of ways. The materials extensively used to cover the ground when cities are built, such as asphalt and concrete, are relatively impermeable. Therefore, when urbanization is extensive, surface runoff tends to be much more concentrated and rapid than before, increasing the risk of flooding. Another problem is that buildings in a floodplain can increase flood heights. The buildings occupy volume that water formerly could fill, and a given discharge then corresponds to a higher stage. Filling in floodplain land for construction similarly decreases the water storage volume available. City storm sewers are installed to keep water from swamping streets during heavy rains, and, often, the storm water is channeled straight into a nearby stream, decreasing the time taken by the water to reach the stream channel and thus increasing the probability of flooding.

Many strategies can reduce flood hazards. For example, accurate flood-hazard mapping allows identification of those areas threatened by floods of different recurrence intervals. Land that is likely to be inundated often might best be restricted to land uses not involving much building, such as for livestock grazing pasture or for parks or other recreational purposes. To some extent, buildings can also be "flood-proofed"—raised on stilts or with lower levels sealed to prevent damage from 100- or 200-year floods.

If open land is available, flood hazards along a stream may be greatly reduced by the use of **retention ponds,** large basins that catch some of the surface runoff, keeping it from flowing immediately into the stream. They may be elaborate artificial structures, old abandoned quarries, or, in the simplest cases, fields dammed by dikes of piled-up soil. The latter option also allows the

Figure 14.24 When insufficiently high levees are overtopped, water is trapped behind them. Several days after this flood on the Kishwaukee river in DeKalb, Illinois, the stage of the river (right) had dropped considerably. Water level behind the levee (left) remained much higher, prolonging flooding and increasing flood damage.

land to be used for other purposes, such as farming, except on those rare occasions of heavy runoff when it is needed as a retention pond. Retention ponds are a relatively inexpensive option and have the added advantage of not altering the character of the stream.

Channelization is a general term for various modifications of the stream channel that increase stream discharge and, hence, the rate at which surplus water is carried away. The channel can be widened or deepened, especially where soil erosion and subsequent sediment deposition in the stream have partially filled in the channel. Alternatively, a stream channel might be re-routed—for example by deliberately cutting off meanders to provide a more direct path for the water flow. Such measures do tend to decrease the flood hazard upstream from where they are carried out. However, by causing more water to flow downstream faster, channelization often increases the likelihood of flooding downstream from the alterations. Often, too, intensive and costly ongoing maintenance of the modified channel is required.

Artificial levees can be built to raise the height of the stream banks. This ancient technique was practiced thousands of years ago on the Nile by the Egyptian pharaohs. Confining the water to the channel, rather than allowing it to flow out into the floodplain, however, effectively shunts it downstream faster during high-discharge events, which may increase flood risks downstream. Another problem is that levees may make people feel so safe about living in the floodplain that development will be far more extensive than if the stream were allowed to flood naturally from time to time. Also, if the levees are overtopped by unexpectedly high water, they then trap water *outside* the stream channel for an extended period (figure 14.24).

Yet another approach to moderating streamflow to prevent or minimize flooding is through the construction of flood-control dams along the stream. Excess water is held behind a dam in a reservoir and may then be released at a controlled rate. As previously mentioned, such an artificial change in stream base level alters channel geometry. If the stream normally carries a high sediment load, silting up of the reservoir will force repeated dredging, which can be expensive and presents the problem of where to dump the dredged sediment.

Summary

A stream is any body of flowing water usually confined within a channel. Streams are a part of the hydrologic cycle, through which all water in the hydrosphere flows.

The size of a stream, as measured by its discharge, is a function of climate, near-surface geology, and the size of the drainage basin from which the water is drawn. The force of the flowing water may cause both downcutting and lateral erosion of banks, including meandering. Meander formation and migration, coupled with deposition of alluvial sediments in and around the stream, contribute to the creation of floodplains. Material is moved by streams in solution, in suspension, or along the streambed. The faster the water flow, the coarser the particles moved, and the greater the capacity of the stream to transport sediment. Alluvium deposited where water velocity decreases may take the form of bars or islands in the channel, overbank deposits (including natural levees) in the floodplain, or alluvial fans and deltas at stream mouths. Over time, most streams tend toward an equilibrium longitudinal profile, with gradient decreasing from source to mouth. When sediment influx and outflow are in balance, the result is a graded stream.

Floods are the stream's response to an input of water too large and/or too rapid for the discharge to be accommodated within the channel. The risk of flooding is a function of climate, soil character, presence or absence of vegetation, and topography. Human activities may unintentionally increase flood hazards by filling in the channel or floodplain during construction, covering soil with impermeable materials, or draining runoff to streams quickly via sewers. Use of flood-frequency records may help to identify areas most at risk, but the changes wrought recently by human activities limit the usefulness of historic records. Strategies used to reduce flood hazards include restrictive zoning; the building of retention ponds, artificial levees, or flood-control dams; and various kinds of channelization. These practices, however, are not without some disadvantages or even risks.

Terms to Remember

alluvial fan	hydrosphere
alluvium	incised meanders
bankfull stage	infiltration
base level	knickpoint
bed load	levees
braided stream	load
capacity	longitudinal
channelization	profile
competence	meanders
crest (flood)	mouth
delta	overbank deposit
dendritic	oxbows
drainage	point bar
discharge	radial drainage
dissolved load	rectangular
downcutting	drainage
drainage basin	recurrence
dynamic	interval
equilibrium	retention pond
floodplain	saltation
flood stage	source
graded stream	stage
gradient	stream
headward	suspended load
erosion	terraces
hydrograph	trellis drainage
hydrologic cycle	tributary stream

Questions for Review

1. Briefly summarize the principal processes and reservoirs of the hydrologic cycle. (You may find a sketch helpful.)
2. What is a stream? A drainage basin?
3. What determines the order of a stream?
4. Describe the kind of geologic situation in which you might expect each of the following to develop: (a) rectangular drainage, (b) radial drainage.
5. What is a stream's discharge, and how is it related to stream capacity and load?
6. Briefly describe the development and migration of meanders. How are meanders and oxbow lakes related?
7. What is a braided stream, and how does braiding develop?
8. Why does a stream commonly deposit a delta at its mouth?
9. Sketch the equilibrium longitudinal profile of a graded stream, indicating base level. How is the profile changed by construction of a dam along the stream?
10. What is a flood? Outline briefly how flood hazards are affected by each of the following: (a) intensity of precipitation, (b) soil type, (c) presence of vegetation, (d) construction that adds extensive impermeable cover.
11. What is a flood-frequency curve? Is the term accurate?
12. Cite and briefly explain three strategies for reducing flood hazards, noting strengths and weaknesses of each.

For Further Thought

1. Make a point of visiting a nearby stream. Walk along the channel, noting meandering, areas where the bank is eroding or point bars are being deposited, and any evidence of human modification of the channel.

2. Investigate the availability of flood-hazard maps for a nearby stream. How current are the maps and the data upon which they are based? Have any efforts been made to reduce the flood hazards? If so, what have they been, what have they cost, and what negative effects, if any, have they had?

Suggestions for Further Reading

Bolt, B. A., W. L. Horn, G. A. Macdonald, and R. F. Scott. 1975. *Geological hazards*. New York: Springer-Verlag. (Chapter 7 surveys causes of floods and discusses flood-hazard mitigation.)

Chin, E. H., J. Skelton, and H. P. Guy. 1975. *The 1975 Mississippi River Basin flood*. U.S. Geological Survey Professional Paper 937.

Knighton, D. 1984. *Fluvial forms and processes*. London: Edward Arnold.

Leopold, L. B., M. G. Wolman, and J. P. Miller. 1964. *Fluvial processes in geomorphology*. San Francisco: W. H. Freeman.

Morisawa, M. 1985. *Rivers*. New York: Longman.

Richards, K. 1987. *River channels*. New York: Blackwell.

Ritter, D. F. 1986. *Process geomorphology*. 2d ed. Dubuque, Iowa: Wm. C. Brown Communications, Inc.

Tank, R. W. 1983. *Environmental geology, text and readings*. 3d ed. New York: Oxford University Press. (The section on floods, pp. 218–77, includes several readings on responses to flood hazards.)

Outline

Coastal Zones and Processes

Crashing surf at Salt Point State Park,
California, is characteristic of a high-energy
coastal zone.
© David Muench

Introduction

Coastal areas vary greatly in character and in the kinds and intensities of geologic processes that occur along them. These areas may be dynamic and rapidly changing under the interaction of land and water, or they may be comparatively stable. In this chapter, we briefly review some of the processes that occur along coasts, examine several different types of coastlines, and consider the impacts that various human activities have on them (and vice versa).

Although the terms *coastline* and *shoreline* are sometimes used interchangeably, the **shoreline** is technically the line made by the water's edge on the land, a small, local feature that is constantly shifting and changing shape. The term **coastline** is properly used to describe the overall geometry of the margin of the land, which encompasses a much larger region of the coast and is a somewhat more permanent feature.

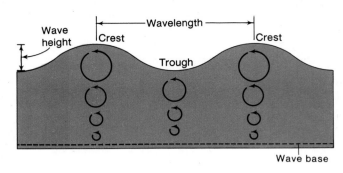

Figure 16.1 Waves and wave terminology. Note the motion of water beneath the surface.

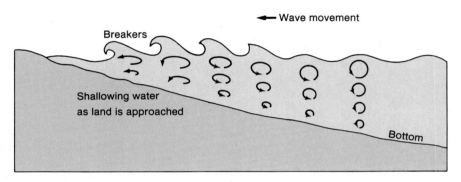

Figure 16.2 Breakers form as undulating waters approaching shore touch bottom, with distortion of water orbits as water shallows.

Waves and Tides

Waves and associated currents are the principal forces behind change along shorelines. Waves, in turn, are induced by the flow of wind across the water surface. Small, local differences in air pressure create undulations in the water surface. The alternating rise and fall of the water surface is a **wave.** If the wind flow continues, the undulations may begin to move laterally across the water surface and even to enlarge.

Waves and Breakers

The shape and apparent motion of a wave reflects the changing form of the water surface; the actual motion of water molecules is different (figure 16.1). In open, deep water, the water moves in circular orbits, relatively large near the surface and decreasing in diameter with increasing depth. The water surface takes on a rippled form. The peak or top of each ripple is a wave

crest; the bottom of each intervening low is a **trough.** The **height** of the waves is simply the difference in elevation between crest and trough, and the **wavelength** is the horizontal distance between adjacent wave crests (or troughs). The **period** of a set of waves is the time interval between the passage of successive wave crests (or troughs) past a fixed point. The **wave base** is the depth at which water motion is negligible; it is approximately half the wavelength.

Only when a wave begins to interact with the bottom does it begin to develop into a *breaker.* This first occurs when the water depth has decreased to about the wave base. Friction with the bottom distorts and ultimately breaks up the orbits, and the water arches over in a breaking wave (figure 16.2). Bottom topography thus plays a major role in the nature of waves reaching the shore. If the coast consists of cliffs and

the water deepens sharply offshore, breaker development may be minimal. The **surf zone** extends outward from the beach to the outermost limit of the occurrence of breakers.

The influence of interaction with the ocean bottom is also seen in the phenomenon of **wave refraction** (figure 16.3). As waves approach a coast, they are slowed first where they first touch bottom, while continuing to move more rapidly elsewhere. As a result, the line of crests is bent, or refracted. If the slope of the bottom is similar all along the coast, the waves are deflected toward projecting points of land, and their energy is concentrated at those points. The refracted waves tend to approach the shoreline more squarely, with wave crests more nearly parallel to the shore, as they move toward it. In recessed bays, where wave energy has been dissipated, sediment deposition may occur.

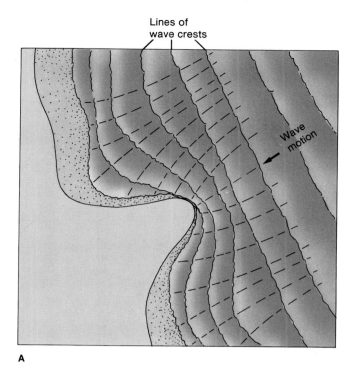

Lines of wave crests

Wave motion

A

B

Figure 16.3 Wave refraction. (A) The energy of waves approaching a jutting point of land is focused on it by refraction. (B) Wave refraction caused by nearshore coral reefs, Oahu, Hawaii.

Waves, Tides, and Surges

As noted previously, waves are localized water-level oscillations set up by wind. **Tides** are broader, regional changes in water level caused primarily by the gravitational pull of sun and moon on the envelope of water surrounding the earth.

If one takes a bucketful of water and swings it quickly in a circle at arm's length, the water stays in the bucket because forces associated with the circular motion push the water outward from the center of the circle (and thus into the bucket). Similarly, as the earth rotates, matter would be flung outward from it, were it not for the restraining influence of gravity. Even so, the earth and the watery envelope of oceans surrounding it show the effects of these rotational forces. The velocity of the moving surface of the earth is greatest near the equator and decreases toward the poles, so earth and water deform to create an equatorial bulge.

Superimposed on this effect of the earth's rotation are the effects of the gravitational pull of the sun and especially the moon. The closer one is to an object, the stronger its gravitational attraction. The moon therefore pulls most strongly on matter on the side of the earth facing it, least strongly on the opposite side. The combined effects of gravity and rotation cause two bulges in the water envelope: one facing the moon, where the moon's gravitational pull on the water is greatest, and one on the opposite side of the earth, where the gravitational pull is weakest and rotational forces dominate (figure 16.4).

As the rotating earth spins through these two bulges of water each day, overall water level at a given point on the surface rises and falls twice daily. This is the phenomenon recognized as tides. The tidal extremes are most significant when sun, moon, and earth are all aligned, and the sun and moon are thus pulling together. The resultant tides are **spring tides** (figure 16.4A). (These have nothing in particular to do with the spring season of the year.) When the sun and moon are pulling at right angles to each other, the difference between high and low tides is minimized. These are **neap tides** (figure 16.4B). The magnitude of water-level fluctuations in any one spot is also controlled, in part, by the underwater topography. The oscillations of waves are superimposed on the tidal regime.

Waves rise and fall in seconds, tides over several hours. A phenomenon of intermediate duration, usually associated with storms, is known as a **surge.**

Figure 16.4 Tides. (*A*) Spring tides: Sun, moon, and earth are all aligned (near times of full and new moons). (*B*) Neap tides: When the moon and the sun are at right angles, tidal extremes are reduced.

It results from some combination of a significant drop in air pressure over an area (which causes a local bulge or rise in water elevation beneath it) and strong onshore winds. Surges from severe storms can easily be several meters high and are most serious when they coincide with the already high spring-tide water levels. On top of that, of course, strong storms can also cause unusually large waves. The combination of surges and high waves can make coastal storms especially devastating, as illustrated by Hurricane Gloria in September 1985 and Hurricane Hugo in 1989.

Beaches and Coastal Features

A **beach** is a gently sloping shore covered by silt, sand, or gravel that is washed by waves and tides. Beaches vary in detail, but certain features are commonly observed.

The Beach Profile

A representative profile or cross section of a beach is shown in figure 16.5. The **beach face** is that portion of the beach exposed to direct overwash of the surf. Its slope varies with the grain size of the beach sediment and the

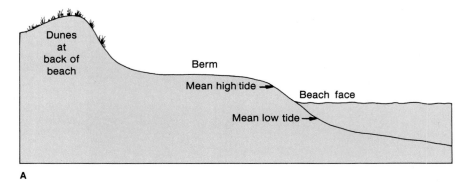

Figure 16.5 A typical beach profile. (*A*) Schematic. (*B*) As illustrated on a Hawaiian beach.

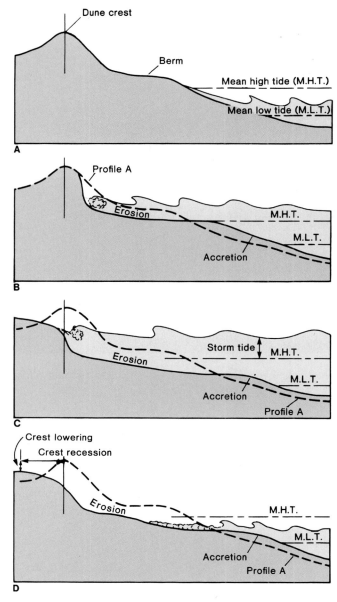

Figure 16.6 Alteration of beach profile due to accelerated erosion by unusually high storm tides. (*A*) Normal wave action. (*B*) Initial attack of storm waves. (*C*) Storm wave attack on foredune. (*D*) After storm wave attack, normal wave action.
Source: "Shore Protection Guidelines," U.S. Army Corps of Engineers.

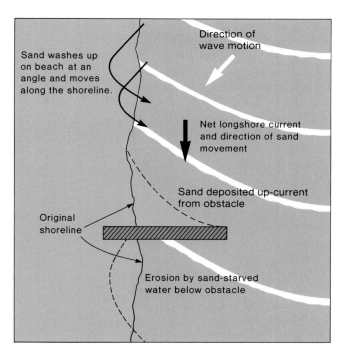

Figure 16.7 Longshore currents and their effect on sand movement. Dashed profile is shoreline after modification.

Sediment Transport at Shorelines

On any beach, the rush of water up the beach face after waves have broken (**swash**) tends to push sediment upslope toward the land. Gravity and the retreating **backwash** together carry it down again. (It is the backwash that is sometimes felt by swimmers as undertow.) A beach profile is described as being in equilibrium when its slope is such that the net upslope and downslope transport of sediment are equal.

If waves approach the beach at an oblique angle, sediment may be moved along the length of the beach, rather than just up and down the beach face perpendicular to the shoreline. Wave refraction tends to deflect waves so that the wave crests approaching the beach are more nearly parallel to the shoreline, but they may still strike the beach face at a small angle. As the water washes up onto and down off the beach, then, it also moves laterally along the coast. This is a **longshore current** (figure 16.7). Likewise, any sand caught up by and moved along with the flowing water is not carried straight up the beach but is transported at an angle to the shoreline. The net result is **littoral**

energy of the waves. To the seaward side of the beach face, there is typically a more shallowly sloping zone, below the low-tide level. Within this zone, sediment may be moved by currents, but it is not actually under attack by waves. Landward of the beach face may be a flat or landward-sloping terrace or **berm** backing up the beach. Not all beaches have berms; some have more than one. The landward limit of the beach can be defined in several ways, depending on the particular situation—for example, by the presence of a rocky cliff or by a zone of permanent vegetation.

Beach profiles are not static. Even a single beach may not always be characterized by the same slope or other features throughout the year. The surges and increased wave action of storms bring about changes that are both sudden and dramatic (figure 16.6). As storms vary seasonally in intensity and frequency, so beach geometry may vary seasonally also.

drift, sand movement along the beach in the same general direction as the motion of the longshore current. Currents tend to move consistently in certain preferred directions on any given beach, which means that, over time, there is continual transport of sand from one end of the beach to the other. On many natural beaches where this occurs, the continued existence of the beach is assured by a fresh supply of sediment produced locally by wave erosion or delivered by streams from farther inland.

Littoral drift is a common and natural beach process. However, beach-front property owners, concerned that their beaches (and perhaps their homes or businesses) will wash away, may erect structures to try to "stabilize" the beach, which generally further alter the beach's geometry (figure 16.7). One common method is the construction of one or more groins or jetties—long, narrow obstacles set more or less perpendicular to the coastline—or break-waters parallel to the coastline (figure 16.8). By disrupting the usual flow and velocity patterns of the currents, these structures change the coastline. Like stream currents, coastal currents slowed by such a barrier tend to drop their load of sand up-current from it. Below (down-current from) the barrier, the water picks up more sediment to replace the lost load, and the beach is eroded. The common result is that a formerly stable, straight shoreline develops an unnatural scalloped shape. The beach is built out in the area up-current of the groins and eroded landward below. Beachfront properties in the eroded zone may be more severely threatened after construction of the "stabilization" structures than they were before.

Any interference with sediment-laden waters can cause redistribution of sand along beachfronts. A marina built out into such waters may cause some deposition of sand around it and, perhaps, within the protected harbor. Farther along the beach, the now-unburdened waters can more readily take on a new sediment load of beach sand. Even modifications far from the coast can affect the beach. One notable

Figure 16.8   Construction of groins and jetties, obstructing longshore currents, causes deposition of sand: Rockaway Beach, New York.
© B. F. Molnia/Terraphotographics

example is the damming of large rivers for flood control, power generation, or other purposes. As indicated in chapter 14, one consequence of the construction of artificial reservoirs is the trapping behind dams of the sediment load carried by the stream. The cutoff of sediment supply to coastal beaches near the mouth of the stream can lead to erosion of the sand-starved beaches. It may be a difficult problem to solve, too, because no readily available alternate supply of sediment might exist near the problem beach areas. Flood-control dams constructed on the Missouri River are believed to be the main reason why the sediment load delivered to the Gulf of Mexico has dropped by more than half over the last thirty-five years. This may explain the recently observed coastal erosion in parts of the Mississippi delta.

When beach erosion is rapid and development (especially tourism) is widespread, efforts have sometimes been made to import replacement sand to maintain wide beaches. The cost is typically very high. Moreover, assuming that the conditions causing erosion still exist, the sand will have to be replenished over and over, at ever-rising cost.

Sometimes, too, when it has not been possible (or has not been thought necessary) to duplicate the mineralogy or grain size of the sand originally lost, the result has been further environmental deterioration. When coarse sands are replaced by finer ones, softer and muddier than the original sand, the finer material more readily stays suspended in the water, clouding it. Off Waikiki Beach in Hawaii and off Miami Beach, delicate coral reef communities have been damaged or killed by this extra water *turbidity* (cloudiness) due to sand replenishment.

Barrier Islands

Barrier islands are long, low, narrow islands paralleling a coastline (figure 16.9). Exactly how or why they form is not known. Some theories suggest that they have formed through the action of longshore currents on delta sands deposited at the mouths of streams.

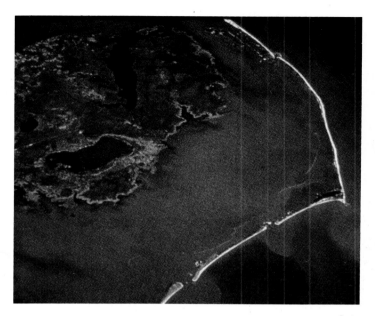

Figure 16.9 Barrier islands along North Carolina's Outer Banks. Photograph by R. Dolan, USGS Photo Library, Denver, Colorado

Figure 16.10 Development on barrier island, Hatteras Island, North Carolina.
© Frank Hanna/Visuals Unlimited

Their formation may also require changes in sea level. However they have formed, they now provide important protection for the water and shore inland from them because they constitute the first line of defense against the fury of high surf from storms at sea. These islands often are so low-lying that water several meters deep may wash completely over them during unusually high storm surges, such as occur during hurricanes, especially along the Gulf and Atlantic coasts.

Because they are usually subject to higher-energy waters on their seaward sides than on their landward sides, most barrier islands retreat landward with time. Typical rates of retreat on the Atlantic coast of the United States are ½ to 2 meters per year, but rates in excess of 20 meters per year have been noted—for example, along some barrier islands in Virginia. Clearly, such settings represent particularly unstable locations in which to build, yet the aesthetic appeal of long beaches has led to extensive development on privately owned sections of barrier islands (see, for example, figure 16.10).

On barrier islands, shoreline-stabilization efforts—building groins and breakwaters and replenishing sand—tend to be especially expensive and, frequently, futile. At best, the benefits are temporary. Construction of artificial stabilization structures is extremely costly; at the same time, it may destroy the natural character of the shore and even the beach, which was the principal attraction for developers in the first place. At least as costly an alternative is to keep moving buildings or rebuilding roads ever farther landward as the beach before them erodes. Expense aside, this is clearly a "solution" for the short term only, and in many cases, there is no place left to which to retreat anyway.

Coastal Erosion

Littoral drift along beaches is not the only cause of coastal erosion. Even rocky cliffs are vulnerable, attacked either by the direct pounding of waves or by the grinding effect of sand, pebbles, and cobbles propelled by waves, which is termed **milling.** Solution and chemical weathering of the rock may cause even more rapid erosion. Because wave action is concentrated at the waterline, above wave base, cliff erosion is often most vigorous there. Also, wave refraction accelerates erosion of projecting points of land.

Figure 16.11 House about to be lost to cliff erosion on the shore of Lake Michigan.
EPA/National Archives

A

B

Figure 16.12 Examples of cliff-protection structures. (*A*) Seawall at Newport, Rhode Island. (*B*) Riprap at base of cliff, El Granada, California. The structure was built in 1973; it may have to be abandoned within decades as erosion continues unabated on either side of the riprap.
(*B*) Photograph courtesy of USGS Photo Library, Denver, Colorado

Sediments are obviously much more readily eroded than are solid rocks, and erosion of sandy cliffs may be especially rapid. Removal of material at and below the waterline undercuts the cliff, leading, in turn, to slumping and sliding of sandy sediments and the swift landward retreat of the shoreline.

Many who build on coastal cliffs are unpleasantly surprised to discover how quickly the coastline can change. Unanticipated and rapid cliff erosion can be an especially dramatic threat (figure 16.11). Sandy cliffs can be cut back by several meters a year.

Various measures that are often as unsuccessful as they are expensive may be tried to halt the erosion. A common practice is to place some type of barrier at the base of the cliff to break the force of wave impact (figure 16.12). The protection may take the form of a solid wall (*seawall*) of concrete or other material, or a pile of large boulders or other blocky debris (*riprap*). If the obstruction is placed only along a short length of cliff directly below a threatened structure, the water is likely to wash in beneath, around, and behind it, rendering it largely ineffective. Erosion continues unabated on either side of the barrier; loss of the protected structure will be delayed but probably not prevented. Waves may also bounce off, or be reflected from, a short length of smooth seawall, to attack a nearby unprotected cliff with greater force.

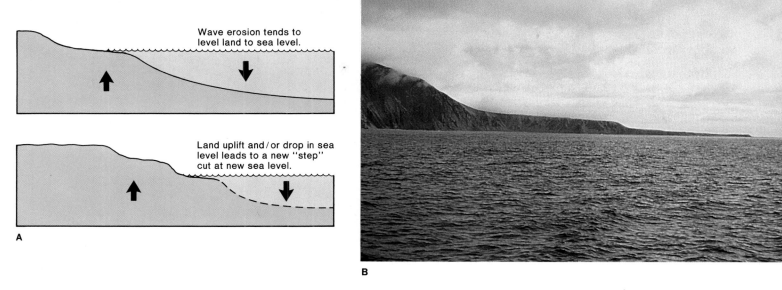

Figure 16.13 Wave-cut platforms. (*A*) Schematic: Wave-cut platforms form when land is elevated or sea level falls. (*B*) Wave-cut platforms of Mikhail Point, Alaska.
(*B*) Photograph by J. P. Schafer, USGS Photo Library, Denver, Colorado

An alternative approach to direct shoreline protection is to erect break-waters farther away from the shore, to reduce the energy of the pounding waves reaching shore. This may slow the erosion but, again, is unlikely to stop it, especially if the cliffs are sandy or made of other weak or unconsolidated materials. Moreover, a breakwater also tends to cause altered patterns of sediment deposition offshore, which may prove inconvenient or may damage marine life.

Emergent and Submergent Coastlines

Water levels vary relative to coastal land, over the short term, as a result of tides and storm surges. Over longer periods, water levels may steadily shift up or down as a result of tectonic processes or from changes related to glaciation. When the coastal land rises or sea level falls, the coastline is described as *emergent*. A *submergent* coastline is found when the land is sinking relative to sea level, or sea level is rising.

Causes of Long-Term Sea-Level Change

As plates move, crumple, and shift, continental margins may be uplifted or dropped down. Such movements can

shift the land by several meters in a matter of seconds to minutes. The results are abrupt, permanent changes in the geometry of the land/water interface and the patterns of erosion and deposition.

In regions overlain and weighted down by massive ice sheets during the last ice age, the lithosphere was downwarped by the load, as noted in chapter 12. Tens of thousands of years later, the lithosphere is still slowly rebounding isostatically to its pre-ice elevation. Where thick ice extended to the sea, a consequence is that the coastline is slowly rising relative to the sea.

Ice caps, as noted in chapter 15, represent an immense reserve of water. As this ice melts, sea levels rise worldwide. Such simultaneous, global changes in sea level are termed **eustatic** changes. (Note that postglacial isostatic rebound partially counters the eustatic sea-level rise resulting from melting ice, in terms of the relative elevation of land and sea.)

Varying rates of plate-tectonic processes can also affect sea level. When seafloor spreading is rapid, there is much young, warm, expanded lithosphere under the oceans, and sea levels rise. When seafloor spreading is very slow, more of the oceanic lithosphere has cooled and contracted, deepening the ocean basins, and sea levels fall.

All of these changes in the relative elevation of land and water may produce distinctive coastal features, which can sometimes be used to identify the kind of vertical movement that is occurring.

Wave-Cut Platforms

Given enough time, wave action tends to erode the land down to the level of the water surface, creating a wave-cut platform at sea level. If the land rises in a series of tectonic shifts and stays at each new elevation for some time before the next movement, each rise results in the erosion of a portion of the coastal land down to the new water level. The eventual product is a series of steplike terraces, called **wave-cut platforms** (figure 16.13). These develop most readily on rocky coasts, rather than on coasts consisting of soft, unconsolidated material. The surface of each such step—each platform—represents an old water-level marker on the continent's edge. Wave-cut platforms can be formed when the continent is rising relative to the sea, or when sea level is falling with respect to the land.

Drowned Valleys and Estuaries

When, on the other hand, sea level rises or the land drops, one result is that streams that once flowed out to sea now have the sea rising partway up the stream valley from the mouth. A portion of the floodplain may be filled by encroaching seawater, forming a **drowned valley** (figure 16.14).

Many drowned valleys have become estuaries. An **estuary** is a body of water along a coastline, open to the sea, in which the tide rises and falls, and which contains a mix of fresh and salt water (*brackish* water). San Francisco Bay, Chesapeake Bay, Long Island Sound, and Puget Sound are examples. Over time, the complex communities of organisms in each estuary have adjusted to the salinity of that particular water. The salinity itself reflects the balance between freshwater input, usually river flow, and salt water. Any modifications that alter this balance change the salinity and can have a catastrophic impact on the organisms. Also, water circulation in estuaries is often very limited. This makes them especially vulnerable to pollution: Because they are not freely flushed out by vigorous water flow, pollutants can accumulate. In this sense, it may be unfortunate that many of the world's large coastal cities are located beside estuaries.

A variation on the drowned valley is produced by the advance and retreat of coastal glaciers. Glaciers flowing off land and into water do not just keep eroding deeper valleys under water until they melt. Ice floats; freshwater glacial ice floats especially readily on denser salt water. Therefore, the carving of glacial valleys essentially stops at the shore. During the last ice age, when sea level worldwide was as much as 100 meters lower than it now is, glaciers at the edges of continental landmasses cut valleys into what are now submerged continental shelf areas. With the retreat of the glaciers and the concurrent rise in sea level, these old glacial valleys were emptied of ice and partially filled with water. That is the

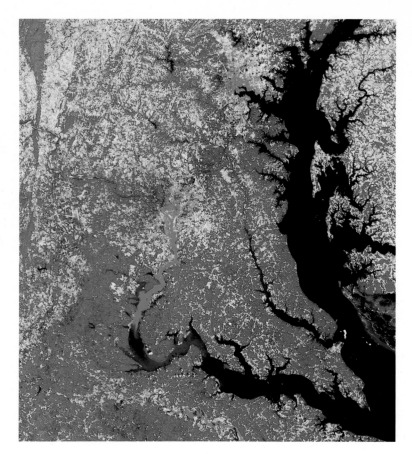

Figure 16.14 Landsat satellite photograph of a drowned valley: Chesapeake Bay.
© NASA

origin of the steep-walled *fjords* so common in Scandinavian countries.

Present and Future Sea-Level Trends

Much of the coastal erosion presently plaguing the United States is the result of gradual but sustained eustatic sea-level rise, probably from the melting of the remaining polar ice. The rise is estimated at about 1/3 meter (1 foot) per century. While this does not sound particularly threatening, two additional factors should be considered. First, many coastal areas slope very gently or are nearly flat, so that a small rise in sea level translates into a far larger inland retreat of the shoreline. Rates of shoreline retreat (landward movement of the shoreline) due to rising sea level have, in fact, been measured at several meters per year in some

low-lying coastal areas. Second, there is concern that global warming will begin to melt remaining ice caps more rapidly, accelerating the sea-level rise (see the discussion of the *greenhouse effect* in chapter 17).

Consistently rising sea levels are a major reason why coastline stabilization efforts repeatedly fail. The problem is not only that the high-energy coastal environment presents difficult engineering challenges (see box 16.1). The problems themselves are intensifying, as shoreline retreat brings water higher and farther inland, pressing ever closer to and more forcefully against more and more structures developed along the coast. Barrier islands and low, exposed beaches are especially vulnerable. It has been estimated that, along an open coastline, a rise in sea level of 1 foot might cause erosion of as much as 100 to 200 feet of beach.

BOX 16.1

Trouble on the Texas Coast: A Case Study

The Texas coast, like much of the Atlantic margin of the United States, is rimmed with barrier islands and barrier peninsulas. The area is subject to severe storms, accompanied by storm surges (commonly several meters or more above normal high tides), strong winds, and high waves. Over the last century, an average of more than one tropical storm or hurricane a year has made landfall somewhere along the Texas coast.

A period of quiet weather in the 1950s was accompanied by a rush of development along the Texas coast. Another period of accelerated building occurred in the 1970s. Building continues today. Meanwhile, the average landward retreat of the beaches progresses at rates of 2 to 7 meters per year. In some especially unstable areas, shoreline changes of over 20 meters per year have been recorded.

As in other dynamic environments, stabilization efforts have had mixed results. In 1900, a particularly fierce hurricane hit Galveston, washing away two-thirds of the city's buildings and causing six hundred deaths. This storm claimed the largest recorded death toll of any historic natural disaster in the United States. In 1902, in response to the devastation of the 1900 hurricane, the Galveston seawall was built. The nearly 6-meter-high structure, built at a cost of $12 million, has offered valuable protection to structures on land, especially during subsequent storms. However, it has also demonstrated some of the permanent changes that can result from seawall construction.

As noted earlier in the text, a portion of wave energy is reflected back from smooth-faced seawalls, so the sand in

Figure 1 The Galveston Seawall.
Photograph by W. T. Lee, USGS Photo Library, Denver, Colorado

front of them is more actively eroded. Longshore currents are commonly strengthened along a seawall, and the seawall cuts off the supply of sand from any dunes at the back of the beach to the area in front of the seawall. The result, typically, is gradual loss of the sandy beach in front of the seawall, within one to fifty years after construction. The beach in front of the Galveston seawall, once several hundred meters wide, is gone now (figure 1). Moreover, the unprotected area at either end of the seawall is being eroded very rapidly. Changes caused by the seawall have made it likely that the seawall will eventually have to be replaced by another, larger, more expensive structure.

Beach replenishment has not often been practiced on the Texas coast, partly for lack of funds, partly for lack of suitable replacement sand. Such an effort was undertaken at Corpus Christi, however, using sand from a nearby river. It has succeeded in the sense that a sandy beach has been maintained there. On the other hand, the river sand is much coarser-grained than was the original beach sand. It has stabilized at a steeper slope angle, both above and below the waterline, than characterized the original beach, and the beach has consequently become less suitable for use by small children. Also, the replenishment efforts must continue if the new beach is not to be eroded away in its turn.

Both of these examples illustrate again two principles: first, that coastal engineering projects permanently change the coastline, and second, that once such engineering efforts are begun, they must be continued unless structures or beaches are ultimately to be abandoned altogether.

Recognition of Coastal Hazards

It is often possible to identify the most unstable or threatened areas along a coast, so as to avoid building on them. The ability to do so depends both on observations of present conditions and on some knowledge of the area's history.

The best setting for building near a beach or on an island, for instance, is at a relatively high elevation (5 meters or more above normal high tide, to be above the reach of storm surges) and in a spot protected by many high dunes between the proposed building site and the water. Thick vegetation, if present, will help to stabilize the beach sand. Information about what has happened in major storms in the past also is very useful. Was the site flooded? Did the overwash cover the whole island? Were protective dunes destroyed?

On either beach or cliff sites, one very important factor is the rate of erosion (if any). Information might be obtained from people who have lived in the area for some time. Better and more reliable guides are old ground or aerial photographs, if available, from the U.S. or state geological survey, county planning office, or other sources. Old, detailed maps that show how the coastline looked in past times can also be used. Knowledge of when the photos were taken or maps made and comparison with the present configuration allows estimation of the rate of erosion. It also should be kept in mind that shoreline retreat in the future may be more rapid than it has been in the past as a consequence of more rapidly rising sea levels. On cliff sites, too, there is landslide potential to consider, and in a seismically active area, the dangers are greatly magnified.

It is advisable to find out what shoreline modifications are in place or are planned, not only close to the site of interest but elsewhere along the coast. These can have impacts in places considerable distances away from where they are actually built. Sometimes, aerial or even satellite photographs make it possible to examine the patterns of sediment distribution and movement along the coast, which should help in the assessment of the likely impact of any shoreline modifications. That such structures exist or are being contemplated may itself be a warning sign! A history of repairs to or rebuilding of structures suggests not only a very active coastline, but also the possibility that, in the future, protection efforts might have to be abandoned for economic reasons.

Summary

Coastal areas are, by nature, dynamic geologic environments, and many are undergoing rapid change. Erosion is a major factor, with waves and currents the principal agents. Waves are generated by the action of wind on the water surface. They develop into breakers as they approach shore, where they may also be deflected by wave refraction in shallowing water. Storms intensify wave erosion, both by increasing wave heights and by causing storm surges that raise overall water levels. In addition to net erosion or deposition of sediment, many beaches are subject to lateral transport of sand by longshore currents. Interference with this process of littoral drift—for example, through construction of piers or jetties—results in redistribution of sediment and altered patterns of sediment erosion and deposition. The same is often true of other shore-protection structures. Even changes far from the shore may affect it, as when trapping of stream sediments by flood-control dams starves coastal beaches of sand and accelerates their disappearance.

An overall rise or fall of the land relative to the sea may result, respectively, in the formation of wave-cut platforms, or drowned valleys and fjords. Many drowned valleys are estuaries, characterized now by a mix of fresh and salt water and by restricted water flow. At present, eustatic sea-level rise, probably caused by ongoing melting of ice caps, is intensifying many coastal problems, especially in low-lying, vulnerable areas, such as barrier islands. Prospective residents of coastal areas can investigate the long-term stability of the coastline through old maps or photographs, in addition to observing present processes active along the coast.

Terms to Remember

backwash
barrier island
beach
beach face
berm
coastline
crest
drowned valley
estuary
eustatic
height (wave)
littoral drift
longshore
  current
milling

neap tides
period
shoreline
spring tides
surf zone
surge
swash
tides
trough
wave
wave base
wave-cut
  platform
wavelength
wave refraction

Questions for Review

1. Sketch a cross section of several waves and indicate the following: wave crests, wavelength, how water is moving beneath the surface.
2. Do breakers typically form in the open ocean? Explain.
3. Briefly describe the origin of tides.
4. Under what circumstances does littoral drift occur? Why does it not necessarily result in complete removal of a beach's sand over time?
5. Describe how a shoreline is altered when groins or breakwaters are erected to slow littoral drift.
6. Cite two possible problems associated with artificial beach-sand replenishment.
7. What is a barrier island? Describe how such islands commonly migrate through time, and comment on the impact of rising world sea levels.
8. What is an emergent coastline? Does it necessarily reflect uplift of the coastal land? Cite one erosional feature characteristic of some emergent coastlines.
9. Describe one cause of eustatic sea-level changes.
10. What is an estuary? Note a situation in which an estuary might develop.
11. Name and describe any three factors that should be considered before undertaking development on a coastal site.

For Further Thought

1. Consider the barrier-island development shown in figure 16.10. Might it have any stabilizing effect on the island? How might you expect the distribution of sand around the island to change over time? What would you expect to be the effect of storm surges?
2. Choose any major coastal city, and investigate the extent of flooding that could be expected in the case of an eustatic sea-level rise of (a) 1 meter and (b) 5 meters. For an example of possible problems and responses to them, investigate what is taking place in the city of Venice, Italy, where a combination of tectonic sinking, surface subsidence, and sea-level rise is causing increasing flooding problems.

Suggestions for Further Reading

American Geological Institute. 1981. Old solutions fail to solve beach problems. *Geotimes,* December, 18–22.

Barnes, R. S. K. ed. 1977. *The coastline.* New York: John Wiley and Sons.

Beer, T. 1983. *Environmental oceanography.* New York: Pergamon Press.

Bird, E. C. F. 1969. *Coasts.* Cambridge, Mass.: M.I.T. Press.

Carter, W. 1988. *Coastal environments.* New York: Academic Press.

Davies, J. L. 1980. *Geographical variation in coastal development.* 2d ed. New York: Longman.

Dolan, R., and H. G. Goodell. 1986. Sinking cities. *American Scientist* 74 (January/February):38–47.

Environmental Protection Agency. 1989. Can our coasts survive more growth? *EPA Journal* 15 (September/October).

Fisher, J. S., and R. Dolan, eds. 1977. *Beach processes and coastal hydrodynamics.* Stroudsburg, Penn.: Dowden, Hutchinson, and Ross.

Heikoff, J. M. 1980. *Marine shoreland resources management.* Ann Arbor, Mich.: Ann Arbor Science.

MacLeish, W. H., ed. 1981. The coast. *Oceanus* 23 (4).

Morton, R. A., O. R. Pilkey, Jr., O. R. Pilkey, Sr., and W. J. Neal. 1983. *Living with the Texas shore.* Durham, N.C.: Duke University Press.

Pethick, J. 1984. *An introduction to coastal geomorphology.* London: Edward Arnold.

U.S. Army Corps of Engineers. 1971. *Shore protection guidelines.* Washington, D.C.: U.S. Government Printing Office.

Walsh, J. J. ed. 1988. *On the nature of continental shelves.* New York: Academic Press.

CHAPTER 17

Outline

Glaciers

Sunlight illuminates an ice cave in the clean,
dense glacial ice of the Bagley Ice Field,
Wrangell-St. Elias National Park, Alaska.
© Marc Muench

Introduction

While ice now covers only about 10 percent of the continental land area, it has sculptured much of the present landscape. Glacial features are found over about three-fourths of the continents' surfaces. Only a few tens of thousands of years ago, sheets of ice covered major portions of North America, Europe, and Asia. Melting glaciers recharged aquifers we use today for water; glacial sediments from older ice advances themselves make up some of the aquifers.

In this chapter, we survey the nature of glaciers, the distinctive erosional and depositional features they produce, and other characteristics of glaciated or cold regions. We also consider how the extent of glaciers has differed in the past (ice ages and their possible causes) and may change in the future (through the alteration of global climate by human activities).

Ice and the Hydrologic Cycle

Approximately 75 percent of the fresh water on earth is stored as ice in glaciers. Water enters this reservoir as precipitation (usually snowfall) and leaves it by evaporation or by melting. The world's supply of glacial ice is the equivalent of sixty years of precipitation over the whole earth, or, to put it another way, represents nine hundred years' flow of all the world's rivers at their present discharge.

In regions where glaciers are large or numerous—in the United States, for example, in the states of Alaska, Washington, Montana, California, and Wyoming—glacial meltwater is an important source of summer streamflow. Anything that modifies glacial melting patterns can, therefore, profoundly affect regional water supplies. Dusting the glacial ice surface with a thin layer of dark material, such as coal dust, increases the heating of the surface and, hence, the rate of melting and water flow. Conversely, cloud seeding over glaciated areas could increase precipitation and the amount of water stored in the glaciers. Increased melt-

water flow can be useful not only in increasing water supplies but also in achieving higher levels of hydroelectric power production. Techniques for modifying glacial meltwater flow are not now being used in the United States, in part because the majority of U.S. glaciers are in national parks, primitive areas, or other protected lands. However, some of these techniques are being practiced in parts of the former Soviet Union and China.

The Nature of Glaciers

Fundamentally, a **glacier** is a mass of ice, on land, formed by recrystallization of snow, that moves under its own weight. A quantity of snow sufficient to form a glacier does not fall in a single winter, so for glaciers to develop, the climate must be cold enough that some snow and ice persist year-round. This, in turn, requires an appropriate combination of elevation and latitude.

Glaciers tend to be associated with the extreme cold of polar regions. However, temperatures also generally decrease at high elevations, so glaciers can exist in mountainous areas even in tropical or subtropical regions. Three mountains in Mexico have glaciers, all at elevations above 5,000 meters (over 16,000 feet). Similarly, there are glaciers on Mount Kilimanjaro in East Africa (which is on the equator), but only at elevations above 4,400 meters (14,500 feet).

Glacier Formation

For glaciers to form, there must be sufficient moisture in the air to provide the necessary precipitation. In addition, the amount of winter snowfall must exceed summer melting so that the snow accumulates year by year. Most glaciers start in the mountains, as snow patches that survive the summer. Slopes that face the poles (north-facing in the northern hemisphere, and vice versa), protected from the strongest sunlight, favor this survival. So do gentle slopes, on which snow can pile up thickly instead of plummeting down periodically in avalanches.

As the snow accumulates, it is gradually transformed into ice. The weight of overlying snow packs it down, drives out much of the air, and causes it to recrystallize into coarser, denser, interlocking ice crystals (figure 17.1). Alternate thawing and freezing after deposition hasten the transformation. (Glacial ice could, in fact, be regarded as a very-low-temperature metamorphic rock.) The material intermediate in texture between freshly fallen snow and compact, solid ice is called **firn.** It is not unlike the coarse frozen material that forms along roadsides in snowy areas during winter and early spring, from piles of snow plowed off the roads. Complete conversion of snow into glacial ice may take from 5 to 3,500 years, depending on such factors as climate and rate of snow accumulation at the top of the pile. Eventually, the mass of ice becomes large enough that it begins to flow outward from the thickest part and, if there is any slope to the terrain, the ice begins to slide or flow downhill.

Movement of Glaciers

The movement of a glacier may be nearly imperceptible, or, for brief periods at least, a glacier may *surge* at up to 10 kilometers per year (about 30 meters, or close to 100 feet, per day). Glacial movements can occur in several ways. The pressure of the ice at its base may be sufficient to melt a little of it, just as ice melts under a skater's blade even in cold weather, and the ice mass as a whole may slide on this meltwater.

Plastic flow and internal deformation within the ice is another mechanism by which glaciers move downslope. This is demonstrated, in part, by the fact that not all portions of a glacier move equally quickly. Typically, flow is fastest at the top and center, in part because friction at the base of the glacier (and the sides of a valley glacier) slows it down (figure 17.2). This is analogous to the effect of friction within the channel on a stream's flow velocity. Also, the internal deformation is most pronounced in the deeper zones of the glacier, which are under greatest con-

Figure 17.1 Progression from fluffy, fresh snow to dense glacial ice.

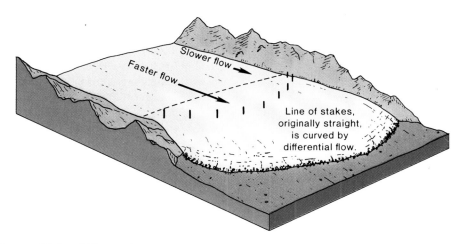

Figure 17.2 Differential rates of movement of different parts of a glacier indicate internal deformation.

fining pressure. The uppermost layers of the glacier, riding on more plastic layers below, behave rigidly or brittlely, fracturing to make the **crevasses,** or cracks, that are the peril of climbers who cross glaciers. Like other rocks, then, glacial ice can develop both folds and faults in response to stress.

The Glacial Budget

Matter is continually cycled through a glacier. Where it is cold enough for snow to fall, fresh material accumulates, adding to the weight of ice and snow that pushes the glacier downhill. Some snow re-evaporates directly in dry, cold air, especially at high latitudes. At some point, too, the advancing edge of the glacier terminates, either because it flows out over water and breaks up, creating icebergs by a process known as **calving** (figure 17.3), or, more commonly, because it has flowed to a place that is warm enough that ice loss (**ablation**) by melting, evaporation, or calving is at least as rapid as the rate at which new ice flows in to replace it (figure 17.4).

Over the course of a year, the size of a glacier varies (figure 17.5). In winter, the rate of accumulation increases, and melting and evaporation decrease. The glacier becomes thicker and also extends farther from its source; it is then said to be *advancing.* In summer, snowfall is reduced or halted, and melting accelerates; ablation exceeds accumulation. The glacier is then described as *retreating,* although, of course, it does not flow backward, uphill. The leading edge simply melts back faster than the glacier moves forward.

Over many years, if the climate remains stable, the glacier achieves a sort of dynamic equilibrium, in which the winter advance just balances the summer retreat, and the average seasonal limits of the glacier remain constant from year to year. An unusually cold (or snowy) period spanning several years would be reflected in a net advance of the glacier over that period, and vice versa. Such phenomena have been observed in recent history. From the mid-1800s to the early 1900s, a

Figure 17.3 Icebergs break free from ice flowing out into water, in the process known as calving.

Photograph by M. Dalechek, courtesy of USGS Photo Library, Denver, Colorado

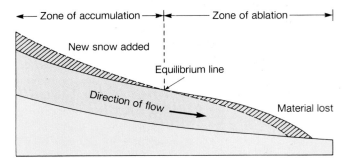

Figure 17.4 Longitudinal cross section of a glacier (schematic). In the zone of accumulation, the addition of new material exceeds loss by melting or evaporation; the reverse is true in the zone of ablation, where there is a net loss of ice.

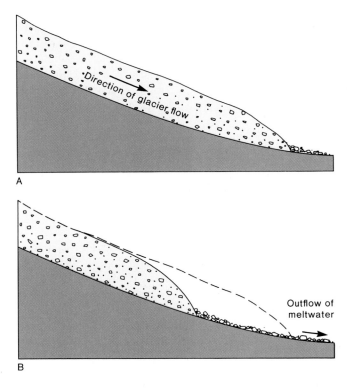

Figure 17.5 (A) Winter advance of glacier carries both ice and sediment downslope. (B) In summer, rapid melting causes apparent retreat of glacier, accompanied by deposition of till from melting ice.

worldwide retreat of many glaciers far up into their valleys was observed. Beginning about 1940, the trend seemed to reverse, and glaciers generally advanced. Glaciers worldwide do not necessarily advance or retreat simultaneously, however, because local climatic conditions may deviate from overall world trends.

Types of Glaciers

Glaciers are divided into two types on the basis of size and occurrence. The most numerous today are the **alpine glaciers,** typically found in mountainous regions, most often at relatively high elevations. Many occupy valleys in the mountains and are called, logically, *valley glaciers* (figure 17.6). Most of the estimated 70,000 to 200,000 glaciers in the world today are alpine glaciers.

The larger and rarer **continental glaciers** are also known as *ice caps* (generally less than 50,000 square kilometers in area) or *ice sheets* (larger). They can cover whole continents and reach thicknesses of a kilometer or more. Though they are far fewer in number than the alpine glaciers, the continental glaciers collectively contain far more ice. At present, the two principal continental glaciers are the Greenland and the Antarctic ice sheets. (The Arctic polar ice mass is not a true glacier, as it is not based on land.) The Antarctic ice sheet is so large that it could easily cover the forty-eight contiguous United States. The geologic record indicates that, at several times in the past, even more extensive ice sheets existed on earth.

Glacial Erosion

The mass and solidity of a glacier make it a very effective agent of both erosion and sediment transport, more so than either wind or liquid water.

Erosional Processes

One way in which glaciers erode rock is through a process known as **plucking** (figure 17.7). Plucking occurs as water seeps into cracked rocks at the base of

Figure 17.6   Alpine glaciers—College Fjord, Alaska. Note the streaks of sediment being carried along by the flowing ice and the tributary glaciers joining the main glacier from the left.
Photograph by W. Montgomery

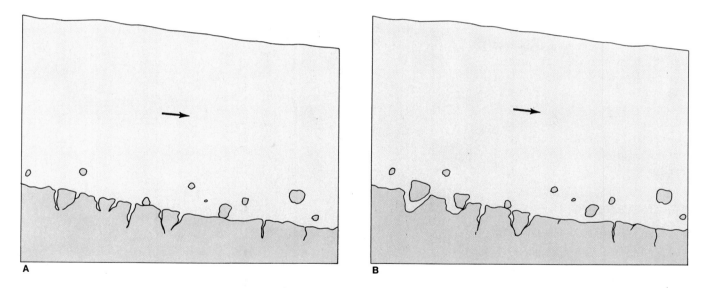

Figure 17.7   Plucking by glacial ice (schematic). (*A*) Water freezes into cracks, attaching rock to glacier. (*B*) Rock fragments frozen into glacier are broken free, plucked out, and transported with the flowing glacier.

the glacier and freezes, attaching the rocks to the glacier. As the glacier moves on, it may tug apart the fractured rock and pluck away chunks of it. The process is accelerated by the fact that water expands as it freezes in the cracks, driving apart rock fragments by **ice wedging.**

Rocks fall from valley walls onto glacial ice and are subsequently carried along with the glacier. Note the dark stripes of sediment along the margins of the alpine glaciers in figure 17.6. Additional material becomes frozen into the ice at the base and sides of the glacier. Ice itself is too soft to erode the

rocks over which it flows, but these rock fragments frozen into the ice cause erosion by **abrasion** of the surrounding rocks at the base of the glacier or along valley walls. The result of glacial abrasion is sediment produced with little alteration of the original chemical and mineralogical character

Figure 17.8 Typical U-shaped glacial valley: Rock Creek Valley, Montana. Note the hanging valley above the main valley floor.

A

of the parent rock. Continued pulverizing into ever-finer fragments eventually produces a powdery, silt-sized sediment termed **rock flour.**

Because ice is solid, it transports sediments with equal efficiency regardless of particle size. Everything is moved together and, when the ice melts, everything is deposited together, as was shown in figure 17.5. The general term for glacially transported sediment is **drift.** Drift may be further subdivided, depending on whether it is stratified or unstratified, as described later in the chapter.

The Glacial Valley

The differences in the character of erosion by water and by ice result in differences in shape between glacial and stream valleys. Glacial valleys are characteristically U-shaped in cross section (figure 17.8). They are broadened and deepened partly by abrasion by rocks frozen into the ice, and partly by plucking. The fjords mentioned in the previous chapter are drowned valleys of glacial origin.

Just as streams may have tributaries, so may alpine glaciers, as valleys join and the ice of several glaciers merges (recall figure 17.6). When the main and tributary glaciers melt, the tributary

B

Figure 17.9 (A) Cirque at head of a valley glacier, Denali National Park, Alaska. (B) Tarns in old glacial cirques, Beartooth Mountains, Montana.

glaciers leave **hanging valleys,** as can be seen in figure 17.8. These smaller, shallower valleys are abruptly truncated by the deeper valley of the main ice mass.

Plucking at the head of an alpine glacier, combined with weathering of the surrounding rocks, produces a rounded or bowl-like depression known as a **cirque** (figure 17.9A). The cirque

Figure 17.10 A horn formed by the erosion of several glaciers around a single peak in Denali National Park, Alaska.

Figure 17.11 An arête formed by two valley glaciers flowing in parallel valleys, Denali National Park, Alaska.

becomes a hospitable setting for more snow accumulation to feed the glacier. When the glacier melts away, a cirque bottom may stay filled with water, making a small, rounded lake called a **tarn** (figure 17.9B).

Other features are formed when more than one glacier erodes a mountain. Where several alpine glaciers flow down in different directions from the same high peak, each chipping away at its head, the result may eventually be a pointed **horn** chiseled from the peak (figure 17.10). Glaciers flowing in parallel, each forming a steep-sided valley, may form an **arête,** a sharp-spined ridge between the valleys (figure 17.11).

Other Erosional Features of Glaciers

Gravel and boulders frozen into the base of the ice and dragged along as the ice moves act as a coarse natural sandpaper. They scrape parallel grooves, or **striations,** in softer rocks over which they move (figure 17.12). Striations are more than evidence of the past presence of a glacier; they also indicate the direction in which the glacier flowed (parallel to the striations). Striations can be produced by both continental and alpine glaciers.

Many alpine glaciers begin by flowing down the valleys of mountain streams. Even a continental glacier may do so locally, widening, deepening, and tending to straighten the stream valley in the process. The Finger Lakes of upstate New York formed in this way, as an ice sheet gouged deeper into old stream valleys. The lakes, too, are elongated in the direction of glacier flow.

The basins now occupied by the Great Lakes were deepened by continental glaciation and filled with glacial meltwater as the glaciers retreated. The basic drainage network of the Mississippi River system was established at the same time, as it carried that meltwater to the sea. Even mountains can be scoured into elongated remnants by the immense mass of ice represented by a continental ice sheet.

Figure 17.12 Striations on a rock surface show the direction of glacial flow. Notice that the grooves extend continuously across fractures. They are clearly a surface feature, not a characteristic of the rock. Devil's Postpile National Monument, California.

Figure 17.13 Till deposited by an Alaskan glacier shows typically poor sorting.

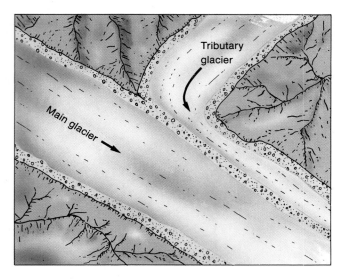

Figure 17.14 Lateral moraines join to become a medial moraine as a tributary glacier joins the main ice mass.

Depositional Features of Glaciers

Sooner or later, melting glacial ice deposits its load of sediment, either directly or through the action of meltwater. This can produce a variety of features. One of the simplest is the isolated large boulder that has been carried along in the ice far from its parent rock. If it cannot be identified as having been derived from the local bedrock, it is a glacial **erratic.**

Direct Deposition: Till and Moraines

Sediment deposited directly by the melting ice is **till.** Till is unstratified and typically poorly sorted with respect to particle size or density, in contrast to most water- or wind-deposited sediments (figure 17.13). Its lithified equivalent is a **tillite.**

A landform made of till is a **moraine.** Moraines come in several forms. The concentrations of rock debris along the sides of a valley glacier, which fall onto the glacier from the valley walls above or are ground out by the ice, are **lateral moraines.** Where tributary glaciers join a valley glacier, the lateral moraines toward the inside join as a ribbon of moraine within the combined ice mass, a **medial moraine** (figure 17.14). Merging of many tributaries eventually gives the resulting composite glacier a striped appearance (recall figure 17.6). When the glacier retreats, recognizable ridges of lateral or medial moraine may be left behind. A broad blanket of ice-deposited till with no particular form is simply **ground moraine.**

Glacial sediment transport continually brings a fresh supply of sediment to the glacier's end. If the extent of the glacier remains the same for several years, a ridge of till accumulates there that is known as an **end moraine.** Formation of a ridgelike landform can also be enhanced by an advancing glacier, acting like a bulldozer on previously deposited ground moraine (figure 17.15). A single glacier may leave multiple end moraines (figure 17.16). In such a case, the one marking the farthest advance of the tongue of ice is the **terminal moraine.** End moraines deposited during glacial retreat, when the ice front was temporarily stationary, are **recessional moraines.**

Outline

Wind and Deserts

Sparse vegetation is a defining characteristic of deserts; the effects of wind are especially pronounced in such a setting because the surface is unprotected by plant growth. White Sands National Monument, New Mexico.
© David C. Schultz

Introduction

Wind is an agent of change, shaping the earth's surface, eroding, transporting, and depositing material, but it is considerably less efficient in that role than is ice or water. On average, worldwide, winds move only a small percentage as much material as do streams. Collectively, mass transport by winds is comparable to mass transport by glaciers, though the latter are areally far more restricted. Wind lacks the ability to attack rocks chemically, by solution, or by wedging during freezing and thawing. Even in many deserts, more sediment is moved during the brief periods of intense surface runoff following occasional rainstorms than by wind during the prolonged dry periods. However, it is in deserts and semiarid lands that the effects of wind action are often most pronounced and most easily observed. Wind erosion can be significant in glacial areas and at high altitudes also.

The Origin of Wind

The flow of streams is driven by the downward pull of gravity. Gravity likewise pulls downward on the atmosphere, but this is not the principal driving force behind the horizontal flow of air at the earth's surface. Air moves over the earth's surface primarily in response to differences in pressure, which commonly correspond to differences in temperature. Local conditions then modify the details of flow direction and speed.

A basic factor in surface temperature variation is latitude. The sun's rays fall most directly, most intensely, on the earth's surface near the equator. Solar radiation is more dispersed near the poles. On a nonrotating earth with a uniform surface, the surface would be heated more near the equator and less near the poles; correspondingly, air over the equator would be warmer than polar air. Warmer, less-dense (lower-pressure) air would rise at the equator, while cooler, denser (higher-pressure) air from the poles would move in toward the lower-pressure region. The rising warm air would spread out lat-

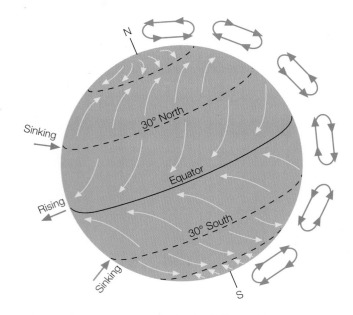

Figure 18.1 Principal present atmospheric circulation patterns.
From Charles C. Plummer and David McGeary, *Physical Geology*, 4th ed. Copyright © 1988 Wm. C. Brown Publishers, Dubuque, Iowa. All Rights Reserved. Reprinted with permission.

erally, cool, and eventually sink. Large circulating air cells would develop, somewhat analogous to mantle convection cells, cycling air from equator to poles and back.

The actual situation is considerably more complicated. For one thing, land and water are heated differentially by sunlight. In addition, surface temperatures over the continents generally fluctuate much more on a daily basis than do temperatures over adjacent oceans. Thus, the irregular distribution of land and water influences the distribution of high- and low-pressure regions, necessarily modifying air flow. Another factor is that the earth rotates on its axis, which adds an east/west component to air movement as viewed from the surface. Also, the earth's surface is not flat; terrain irregularities introduce further complexities in air circulation. Friction between moving air masses and the surface alters both wind direction and wind speed. For example, the presence of tall, dense vegetation can reduce near-surface wind speeds by 30 to 40 percent over the vegetated areas.

A generalized view of large-scale global air-circulation patterns is shown in figure 18.1. Different latitude belts are characterized by different pre-

vailing wind directions. Most of the United States is in a zone of westerlies, in which winds generally blow from west/southwest to east/northeast. Local weather conditions and geography, of course, produce regional deviations from this pattern on a day-to-day basis.

Wind Erosion and Sediment Transport

Flowing air and flowing water have much in common as agents of sediment transport. Both can move particles by rolling them along the surface, by saltation, or in suspension. Both also transport coarser and denser material the faster they move. Water, being denser and more viscous, is more efficient at transporting quantities of material than is wind, but the processes involved are quite similar.

Like water, wind erodes sediment more readily than solid rock. In fact, wind alone has little impact on rock. However, wind-transported sediment can wear away at rock by the process of **abrasion.** Wind abrasion is a sort of natural sandblasting, very similar to milling by sand-laden waves or to glacial abrasion. If winds blow consistently from one or a few directions,

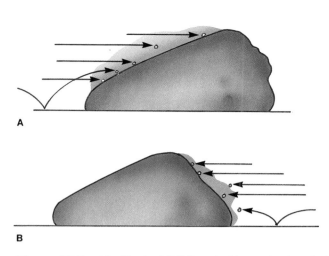

Figure 18.2 Ventifacts. (*A*) If the wind is predominantly from one direction, rocks are planed off or flattened on the upwind side. (*B*) With a persistent shift in wind direction, additional facets are cut in the rock. (*C*) Examples of ventifacts.

(*C*) Photograph by W. N. Lockwood, courtesy of USGS Photo Library, Denver, Colorado

Figure 18.3 (*A*) Undercutting of granite boulder by near-surface wind abrasion. (*B*) Undercut sedimentary rock, Garden of the Gods, Colorado.

(*A*) Photograph by K. Segerstrom, courtesy of USGS Photo Library, Denver, Colorado

exposed cobbles and boulders may be planed off where they face the wind, in time taking on a faceted shape (figure 18.2). Grooves may also be cut in the direction of wind flow. If wind speed is too low to lift the transported sediment very high above the ground, tall rocks may show undercutting close to ground level (figure 18.3). Rocks

sculptured by abrasion are called **ventifacts**—literally, "wind-made" rocks (from the Latin).

Wholesale removal of unconsolidated sediment by wind action is **deflation.** Deflation is naturally most active where winds are unobstructed and the sediment exposed, unprotected by vegetative or artificial cover: deserts,

beaches, unplanted farmland. The key role of vegetation in retarding wind erosion of sediment and soil was demonstrated especially dramatically in the United States in the early twentieth century, during the Dust Bowl era (see box 18.1). A common feature produced by deflation is the descriptively

BOX 18.1

The Dust Bowl

The Dust Bowl area proper, although never exactly defined, comprises close to 100 million acres of southeastern Colorado, northeastern New Mexico, western Kansas, and the Texas and Oklahoma panhandles. After the Civil War, farmers migrating westward found flat or gently rolling land, much of it covered by prairie grasses and wildflowers rather than thick forests, so it was easy to clear and adapt for farming. Over much of the area, native vegetation was removed and the land plowed and planted to seasonal crops. Elsewhere, grazing livestock cropped the prairie. In the 1930s, several years of drought killed the crops, which were not as well adapted to dry conditions as the native prairie vegetation had been. Once the crops died, there was nothing left to hold down the soil and protect it from the west winds sweeping unobstructed across the plains.

The action of the wind was most dramatic during the fierce dust storms, which began in 1932. They were described as "black blizzards" that blotted out the sun (figure 1). Black rain fell in New York State, black snow in Vermont, as windblown dust spread east. People choked on the dust, some dying of suffocation or of a "dust pneumonia" similar to the silicosis miners develop from breathing rock dust.

By the late 1930s, concerted efforts by individuals and state and federal govern-

Figure 1 Example of a 1930s dust storm in the Dust Bowl.
Photograph courtesy of U.S.D.A. Soil Conservation Service

ment agencies to improve farming practices so as to reduce wind erosion—together with a return to more normal rainfall—had considerably reduced the problems. However, drought struck again in the 1950s, and tens of millions of acres of cropland were damaged by wind erosion in 1954. Millions of acres more were damaged in the mid-1970s. In recent years, the

extent of wind damage has been limited somewhat by the widespread use of irrigation to maintain crops in dry areas and dry times. However, as noted in chapter 15, some of the important sources of that irrigation water are rapidly becoming depleted. Future spells of combined drought and wind may yet produce more scenes like those of figure 1.

---

named **blowout,** a bowl-shaped depression hollowed out of a loose, sandy surface.

In some dry areas, the selective removal of finer sediments by wind, often assisted by seasonal surface runoff, produces a surface sediment of residual coarser material. This **desert pavement** effectively protects underlying finer sediments from further erosion (figure 18.4). A desert pavement surface, once established, can be very stable. If the protective layer of coarser gravel and boulders is disturbed, how-

ever, the newly exposed fine sediment may be subject to rapid wind erosion.

Eolian Deposits

Wind-related processes and products are also termed **eolian** (or, in older literature, *aeolian*; named for Aeolus, Latin god of the winds). The relatively few kinds of eolian sedimentary deposits are commonly well sorted. As with water-laid sediments, this sorting results because the maximum particle size moved depends on flow velocity.

Winds selectively move finer, lighter materials; as they slow, they drop coarser, denser grains first.

Dunes

The principal eolian depositional landform is the **dune,** a low mound or ridge of sediment. When they hear the word dune, most people think *sand dune.* However, dunes can be formed of sediment of different sizes, and even of snow (figure 18.5). Dunes begin to form when sediment-bearing winds en-

Figure 18.4 Effects of deflation. Example of desert pavement, formed by selective removal of finer sediments.
Photograph by J. R. Hunt, USGS Photo Library, Denver, Colorado

A

B

Figure 18.5 Examples of dunes. (*A*) Sand dune, coastal Rhode Island. Note the ripples, like miniature dunes, on the dune surface. (*B*) Snow dunes on a barren midwestern field in winter. Note the ground bared by deflation between the dunes.

counter an obstacle that slows them down. With reduced velocity, the wind begins to drop the coarsest, heaviest fraction of its load. The deposition, in turn, creates a larger obstacle that constitutes more of a windbreak, causing more deposition, in a self-reinforcing cycle. Once started, a dune can grow very large. Typical dunes range from 3 to 100 meters in height, but dunes 200 or more meters high are occasionally found. What ultimately limits a dune's size is not known.

Dune Migration

A dune assumes a characteristic profile in cross section (cut parallel to the direction of wind flow): gently sloping on the windward side (the side facing the wind), steeper on the downwind side. With continued wind action, particles are rolled or moved by saltation up the shallower upwind slope and tumble down the steeper face, or **slip face.** The slope of the slip face is characteristic of the sediment of the dune; for sand, it is commonly 30 to 35 degrees from the horizontal. Layering often develops on

Figure 18.6 Erosion has brought out the eolian cross-beds in this outcrop of Navajo Sandstone. One can almost imagine the blowing wind shaping these now-lithified dune sands.
© Wm. C. Brown Communications, Inc./Photograph by Doug Sherman

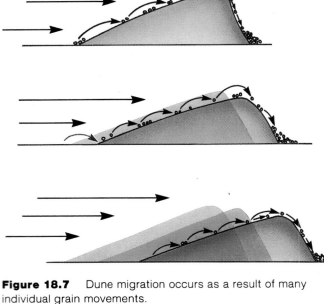

Figure 18.7 Dune migration occurs as a result of many individual grain movements.

Figure 18.8 Marching sand dunes encroaching on trees, Oregon Dunes National Recreation Area.

Figure 18.9 Transverse dunes: crescent-shaped barchan dunes. (See also figure 18.5B.)
Photograph by G. K. Gilbert, courtesy of USGS Photo Library, Denver, Colorado

the slip face. If winds and depositional patterns shift, eolian cross-beds result (figure 18.6).

The net effect of the many individual grain movements under sustained winds from one direction is that the dune itself is moved slowly downwind (figure 18.7). Migrating dunes—especially large ones—can be a real menace where they march across roads, through forests, and even over buildings (figure 18.8). The costs to clear and maintain roads along sandy beaches and through deserts can be high, for dunes can move several meters or more in a year. The usual approach to dune stabilization is to plant vegetation. However, many dunes exist where they do, in part, because the climate is too dry to support vegetation. Aside from any water limitations, young plants may be difficult to establish in shifting dune sands because their tiny roots cannot secure a hold.

Dune Forms

There are several types of dune forms. In a given area, one type will predominate, depending on the particular balance among sediment supply, wind characteristics, and abundance of vegetation.

Transverse dunes are elongated perpendicular to the prevailing wind direction. Many of these have a crescent shape, with arms or "horns" pointing downwind (in the direction of the slip face); these are **barchan dunes** (figure 18.9). Very long, narrow barchan dunes grade into continuous *transverse ridges*, with the same cross-sectional profile but not appearing as discrete mounds of sand. Transverse dunes are common where sediment supply is abundant.

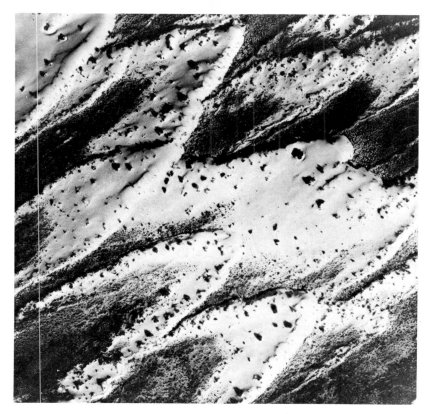

Figure 18.10 Parabolic dunes, with the arms of the dunes anchored by vegetation.
Photograph by E. D. McKee, courtesy of USGS Photo Library, Denver, Colorado

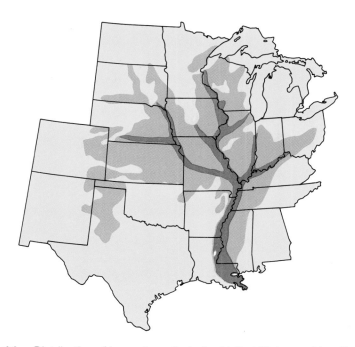

Figure 18.11 Distribution of loess deposits in the United States and location of major glacier-supplied stream valleys.
Source: After J. Thorp and H. T. U. Smith, "Pleistocene Eolian Deposits of the United States, Alaska, and Parts of Canada," Geological Society of America map, 1952.

Longitudinal dunes occur where sediment supply is limited and winds are relatively strong. These dunes are elongated parallel to the direction of wind flow. They may form as the limited quantity of sediment is strung out gradually by the wind.

Where vegetation is more abundant, though not so plentiful as to prevent dune formation altogether, **parabolic dunes** tend to form (figure 18.10). At first glance, they appear similar in shape to barchan dunes, but the arms of parabolic dunes point *upwind*, while the slip face is, as always, downwind. This appears to result from vegetation anchoring the arms of the dune. Wind velocity slows around the vegetation, allowing the vegetation to hold sediment in place, while the bulk of the dune marches on.

Loess

Rarely is the wind strong enough to move sand-sized or larger particles very far or very rapidly. Fine dust, on the other hand, is more easily suspended in the wind and can be carried many kilometers before it is dropped. A deposit of windblown silt is known as **loess.** The rock and mineral fragments in loess are in the range of 0.01 to 0.06 millimeter (0.0004 to 0.0024 inch) in diameter.

The principal loess deposits in the United States are in the central part of the country, and their spatial distribution provides a clue as to their source (figure 18.11). They are concentrated around the Mississippi River drainage basin, particularly on the east sides of major rivers of that basin. Those same rivers drained away much of the meltwater from retreating ice sheets in the last ice age. Glacially produced sediment was washed down and deposited along the river valleys, and the lightest material, much of it rock flour, was blown farther eastward by the prevailing west winds.

Not all loess deposits are ultimately of glacial origin. Some in the western United States may have formed from dust blown off the southwestern deserts. Loess derived from the Gobi Desert covers large areas of China, and

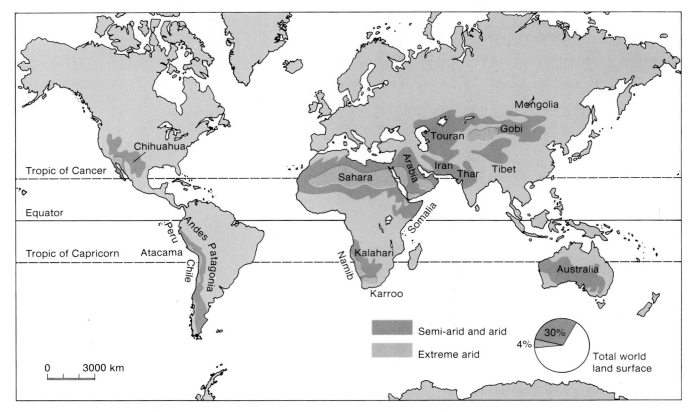

Figure 18.12 Distribution of the world's arid lands.

From A. Goudie and J. Wilkinson, *The Warm Desert Environment.* Copyright © 1977 Cambridge University Press. Reprinted by permission.

additional loess deposits are found downwind of other major deserts. Loess may also be derived from the finest fractions of volcanic ash deposits.

> **B**ecause dry glacial erosion does not involve as much chemical breakdown as does stream erosion, as noted in chapter 17, many soluble minerals are preserved in glacial rock flour. These minerals provide some valuable plant nutrients to the farmland soils now developed on the loess. Newly deposited loess is also quite porous and open in structure, so it has good moisture-holding capacity. These two characteristics together make the farmlands developed on midwestern loess particularly productive.

Loess does have drawbacks with respect to applications other than farming. While its light, open structure is reasonably strong when dry and not heavily loaded, it may not make suit-able foundation material. Loess is subject to *hydrocompaction,* meaning that, when wetted, it tends to settle, crack, and become denser and more consolidated, to the detriment of any structures built on top of it. The very weight of a large building can also cause settling and collapse of loess.

Deserts

A **desert** is a region with so little vegetation that no significant population can be supported on that land. It need not be hot or even, technically, dry. Polar ice caps are a kind of desert. In more temperate climates, deserts are characterized by very little precipitation—25 centimeters (10 inches) per year or less—but they may be consistently hot, cold, or variable in temperature with the season or time of day, depending on latitude and local setting. The distribution of the arid regions of the world (exclusive of polar deserts) is shown in figure 18.12.

Causes of Natural Deserts

A variety of factors contribute to the formation of a desert. One is moderately high surface temperatures. Most vegetation, under such conditions, requires abundant rainfall and/or slow evaporation of what precipitation does fall. The availability of precipitation is governed, in part, by the global air circulation patterns shown in figure 18.1. Because the oceans are the major source of moisture in the air, simple distance from the ocean (in the direction of air movement) can be a factor contributing to the formation of a desert. The longer an air mass is in transit over dry land, the greater chance it has of losing some of its moisture through precipitation.

Warm air holds more moisture than cold. Similarly, when the pressure on a mass of air is increased, the air can hold more moisture. Air spreading outward from the equator at high altitudes is chilled and at low pressure, since air pressure and temperature decrease

Outline

Mass Movement, Mass Wasting

Despite obvious potential for landslides,
development sometimes occurs in high-risk
settings, such as at the base of Cantagalo
Rock, Rio de Janeiro, Brazil.

Photograph by F. O. Jones, USGS Photo Library,
Denver, Colorado

Introduction

While the internal heat of the earth drives mountain-building processes, just as inevitably, the force of gravity acts to tear the mountains down. Gravity is the great leveler. It pulls constantly downward on every mass of material everywhere on earth, causing a variety of phenomena collectively termed **mass wasting,** whereby geological materials are moved downslope from one place to another without a transporting agent, such as wind or water. Erosion is one form of mass wasting. The movement associated with mass wasting can be slow, subtle, almost undetectable on a day-to-day basis; or that movement can be sudden, as in a rockslide or avalanche, and the swift devastation obvious. The term **"landslide"** is a very general term used to describe rapid mass movements.

In the United States alone, landslides and related slower mass movements cause over $1 billion in property damage every year, though losses of human life are relatively small. Many mass movements occur quite independently of human activities. In some areas, active steps have been taken to control downslope movement or to limit the resultant damage. On the other hand, certain human activities aggravate local landslide dangers, usually as a result of failure to take those hazards into account. Large areas of this country—and not necessarily only mountainous regions—are potentially at risk from landslides (figure 19.1).

Causes and Consequences of Mass Movements

Basically, mass movements occur whenever the downslope pull due to gravity overcomes the forces resisting it. Sudden movements may be set off by a triggering mechanism. Mass movements, in turn, may cause secondary problems, such as flooding, as we will see later in the chapter.

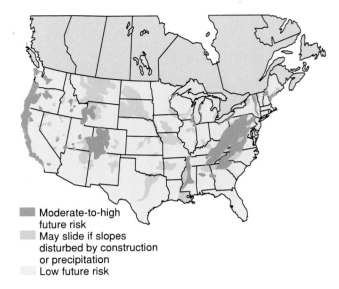

Figure 19.1 Landslide hazards in the United States.
Source: *U.S. Geological Survey Professional Paper 950.*

Moderate-to-high future risk
May slide if slopes disturbed by construction or precipitation
Low future risk

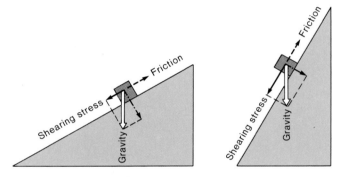

Figure 19.2 Effects of slope geometry on slide potential. The mass of the block, and thus the total downward pull of gravity, is the same in both cases, but the steeper the slope, the greater the shearing-stress component.

Effects of Slope

Steepness of slope is one major factor contributing to the probability of a landslide: All else being equal, the steeper the slope, the greater the downslope pull, or **shearing stress** (see figure 19.2). Counteracting the shearing stress is *friction,* in the case of unconsolidated material, or **shear strength,** in the case of solid rock. When shearing stress exceeds frictional resistance or the shear strength of the material, sliding occurs.

For dry, unconsolidated material, the **angle of repose** is the maximum slope angle at which the material is stable (figure 19.3). This angle varies with the material. Smooth, rounded particles tend to support only very low-angle slopes (imagine making a heap of marbles or ball bearings), while rough, sticky, or irregular particles can be piled more steeply without becoming unstable.

Solid rock can be perfectly stable even at a vertical slope but may lose its strength if it is broken up by weath-

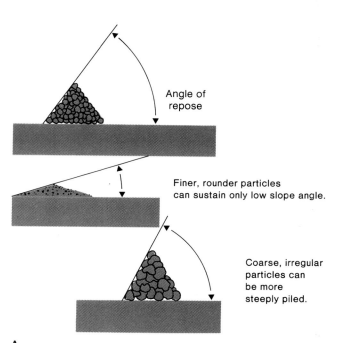

Figure 19.3 Angle of repose, an indicator of an unconsolidated material's resistance to sliding. (*A*) Coarser and rougher particles can maintain steeper stable slope angles. (*B*) This steep cliff is not truly stable; eventually the soil collapses to assume its natural angle of repose. (*B*) Photograph by J. T. McGill, USGS Photo Library, Denver, Colorado

Labels in figure A: Angle of repose; Finer, rounder particles can sustain only low slope angle. Coarse, irregular particles can be more steeply piled.

ering or fracturing. Also, in layered sedimentary rocks, there may be weakness along bedding planes, where different rock units are imperfectly held together; or some units may themselves be weak or even slippery (clay-rich layers, for example). Such planes of weakness are potential slide planes.

Slopes may be steepened to unstable angles by natural erosion by water or ice. Erosion also can undercut rock or soil, removing the support beneath a mass of material and thus leaving it susceptible to falling or sliding. This is a common contributing factor to landslides in coastal areas or along stream channels. Over long pe-

riods of time, slow tectonic deformation can also steepen the angles of slopes or bedding planes, pushing them to instability.

Effects of Fluid

Aside from its role in erosion, water can greatly increase the likelihood of mass movements in other ways. For example, it can seep along bedding planes in layered rock, reducing friction and making sliding more likely. As noted in connection with earthquakes and faulting, an increase in pore water pressure in saturated rocks decreases

the rocks' resistance to shearing stress, which can also facilitate sliding.

In unconsolidated materials, the role of water is variable. A little moisture may add some cohesion (it takes damp sand to build a sand castle). However, substantial increases in water content both increase pore pressure and reduce the friction between particles that provides strength. The very mass of water in saturated soil may add enough extra weight, enough additional downward pull, to set off a landslide on a slope that was stable when dry (see figure 19.4).

Frost wedging and the associated fracturing can weaken rocks. Some soils rich in clays absorb water readily; one type of clay, montmorillonite, may absorb twenty times its weight in water and form a weak gel. Such material fails easily under stress. Other clays expand when wet, contract when dry, and can destabilize a slope in the process. In terms of property damage, expansive clays are, in fact, the most costly geologic hazard in the United States; see figure 19.5.

Earthquakes

Landslides are a common consequence of earthquakes in hilly terrain, as noted in chapter 10. Seismic waves passing through rock stress and fracture it. The added stress may be as much as half that already present due to gravity. Ground shaking also jars apart soil particles and rock masses, reducing the friction that holds them in place.

One of the most lethal earthquake-induced landslides occurred in Peru in 1970. There had already been an earlier slide below the steep, snowy slopes of Nevados Huascarán, the highest peak in the Peruvian Andes, in 1962. This slide, which was not triggered by an earthquake, had killed approximately 3,500 people. In 1970, a magnitude 7.7 earthquake centered 130 kilometers to the west shook loose a much larger debris avalanche that buried most of the towns of Yungay and Ranrahirca and more than 18,000 people with them. Some of the debris was estimated to have moved at 1,000 kilometers per hour. In the 1964 Alaskan earthquake, the Turnagain Heights section of Anchorage also was heavily damaged by landslides.

Figure 19.4 Slope failure under road after rains, Floresta Creek Valley, Brazil.
Photograph by F. O. Jones, USGS Photo Library, Denver, Colorado

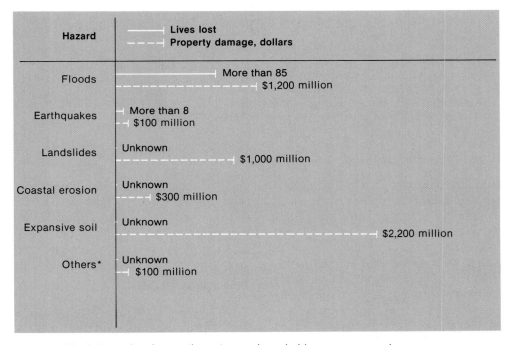

*Includes volcanic eruptions, tsunamis, subsidence, creep, and other phenomena.

Figure 19.5 Relative magnitude of annual loss of life and property damage in the United States from various geologic hazards.
Source: *U.S. Geological Survey Professional Paper 950.*

Figure 19.6 An old 3-million-cubic-meter slide near Thistle, Utah, was reactivated in the wet spring of 1983. It blocked Spanish Fork Canyon and cut off highway and rail routes.

Photograph courtesy of USGS Photo Library, Denver, Colorado

Quick Clays

A geologic factor that contributed to the 1964 Anchorage landslides and that continues to add to the landslide hazards in Alaska, California, parts of northern Europe, and elsewhere is a material known as **quick clay** or **sensitive clay**. True *quick clays* are most common in northern latitudes. They are formed from glacial rock flour deposited in a marine environment. When this extremely fine sediment is later uplifted above sea level by tectonic movements, it contains salty pore water. The sodium chloride in the pore water acts as a "glue," holding the clay

particles together. Fresh water subsequently infiltrating the clay washes out the salts, leaving a delicate, honeycomblike structure of particles. Vibration from seismic waves breaks the structure apart, reducing the quick clay to a fraction of its original strength, creating a finer-grained equivalent of quicksand that is highly prone to sliding.

So-called sensitive clays are similar in behavior to quick clays, but differently formed. Weathering of volcanic ash, for example, can produce a sensitive clay sediment, *bentonite*. Such deposits are locally common in the western United States, where there has been much relatively recent volcanism.

Effects of Vegetation

Vegetation tends to stabilize slopes. Plant roots, especially those of larger shrubs and trees, can provide a strong interlocking network to hold unconsolidated materials together and prevent flow. (This benefit may continue for several years after trees are cut down, before the roots decompose.) Actively growing vegetation also takes up moisture from the upper layers of soil and can thus reduce the overall moisture content of the mass, increasing its shear strength.

Secondary Consequences of Mass Movements

Just as landslides can be a result of earthquakes, other events, notably floods, can be produced by landslides. A stream in the process of cutting a valley may create unstable slopes. Subsequent landslides into the valley can dam up the stream, creating a natural reservoir. The filling of the reservoir makes the area behind the earth dam uninhabitable, though this usually happens slowly enough that lives are not lost (see figure 19.6).

A further danger is that the unplanned dam formed by the landslide may later fail. In 1925, an enormous rockslide a kilometer wide and over 3 kilometers long in the valley of the Gros Ventre in Wyoming blocked that valley, and the resulting lake stretched to 9 kilometers long. After spring rains and snow melt, the "dam" failed. Flood waters swept down the valley below. Six people died, and the toll would have been far higher in a more populous area.

Landslides in and near dams and reservoirs can also cause problems. In the reservoir disaster at Vaiont, Italy, the reservoir aggravated the local landslide hazards, and the subsequent slides and floods destroyed several towns; see box 19.1.

BOX 19.1

An Anthropogenic Landslide: The Vaiont Dam Disaster

The Vaiont River flows through an old glacial valley in the Italian Alps. The valley is underlain principally by limestones with some clay-rich layers, all folded and fractured during the building of the Alps. The beds on either side of the valley dip down toward the valley floor (figure 1). The rocks are particularly prone to sliding along the clay-rich layers. Abundant evidence of old rockslides can be seen in the valley.

In 1960, the Vaiont Dam, built for power generation, was completed. As the water level in the reservoir behind the dam rose, pore pressures of groundwater in the rocks of the reservoir walls rose also, and the clays of the clay-rich layers swelled. During 1960, a block of 700,000 cubic meters of rock slid from the slopes of Monte Toc, on the south wall, into the reservoir. Creep was noted over a still greater area. Through 1960–1962, measured creep rates occasionally reached 25 to 30 centimeters (nearly 1 foot) per week.

Late summer and fall of 1963 were times of heavy rainfall in the Vaiont valley. The saturated rocks represented more mass pushing downward on zones of weakness. Groundwater flow increased. The water table and reservoir level rose by over 20 meters. Creep rates increased. On October 1, animals that had been grazing on the slopes of Monte Toc moved off the hillside and would not return.

The rains continued, and so did the creep. On October 8, the engineers tried to reduce the water level in the reservoir by opening the gates of two outlet tunnels. Yet the water level continued to rise: The silently creeping mass had begun to encroach on the reservoir, reducing its volume. By October 9, creep rates were as high as 80 centimeters per day.

At about 10:40 on the night of October 9, in the midst of another downpour, dis-

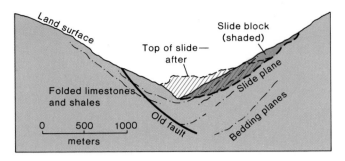

Figure 1 Cross section of the Vaiont River Valley.

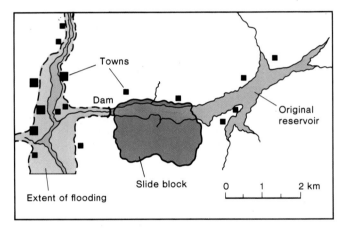

Figure 2 The Vaiont Reservoir slide block. Note areas and towns inundated by the resulting flooding.

From G. A. Kiersch, "The Vaiont Reservoir Disaster," *Mineral Information Service* 18(7):129–138. Reprinted by permission of the California Division of Mines and Geology.

aster struck. A 240–million-cubic-meter chunk of hillside slid into the reservoir (figure 2). The shock of the slide was detected on seismometers throughout Europe. The displaced water crashed over the dam and rushed down the valley below in a wall 100 meters high. Within a period of about five minutes, nearly 3,000 people were drowned and entire towns obliterated.

It is a great tribute to the designer of the Vaiont Dam that the dam itself held, resisting forces far beyond its design specifications. However, those who chose the dam site had every reason to realize the landslide risks since there was ample evidence of persistent slope instability. In fact, the dam builders in this instance were later found guilty of gross negligence and imprisoned.

Figure 19.7   Rockfall (schematic).

Figure 19.8   Accumulation of talus from multiple rockfalls, Devil's Postpile National Monument.

Figure 19.9   Rockslide (schematic): A coherent mass of rock moves as a unit when inadequately supported.

Types of Mass Movements

Mass movements can be subdivided on the basis of the type of material moved and the nature of the movement, including the rate of movement. The material moved can be either unconsolidated, fairly fine material (for example, soil or snow) or large, solid blocks or sheets of rock. It can be quite uniform in size or a mix of fine sediment and boulders. A few examples may clarify the different types of mass movements.

Differences in Materials and Style of Motion

A **fall** is a free-falling action in which the moving material is not always in contact with the ground below. Falls are most often *rockfalls* (figure 19.7). They frequently occur on very steep slopes or cliffs, when rocks high on the slope, weakened and broken up by weathering, lose support as materials under them are eroded away. Repeated rockfalls in one place over a period of time can result in the accumulation of piles of blocky debris, termed **talus** (figure 19.8).

In a **slide,** a fairly coherent unit of rock or soil slips downward along a clearly defined surface or plane. Rockslides most often involve movement along a bedding plane between successive layers of sedimentary rocks, or slippage where differences in the original composition of the sedimentary layers result in a weakened layer or a surface with little cohesion (figure 19.9).

If the rock or soil mass moves over only a short distance, the slide may be termed a **slump.** Slumps in soil, rather than rock, are often characterized by a rotational movement that accompanies the downslope movement of the soil mass. The surface at the top of the slide may be relatively undisturbed. The cliff-like scar at the head of the slump is a type of **scarp.** The lower part of the slump block may end in a flow (figure 19.10).

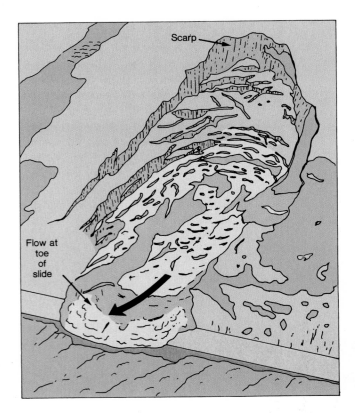

Figure 19.10 Slump in soil (schematic): Most of the soil mass moves coherently, and the surface is disturbed relatively little.

Figure 19.11 Debris avalanche. (*A*) Schematic: The moving mass is a chaotic jumble of varied materials. (*B*) Puget Peak avalanche following 1964 Alaskan earthquake.

(*B*) Photograph by G. Plafleer, USGS Photo Library, Denver, Colorado

Landslides involving unconsolidated, noncohesive material are extremely common. These are **flows,** in which material moves in a chaotic, disorganized fashion, with mixing of particles within the flowing mass, as a fluid moves. Flows need not involve only soils. Snow avalanches are one kind of flow; nuées ardentes are another, associated only with volcanic activity (see chapter 4). Where soil is the flowing material, flow phenomena may be described as *earthflows* (fairly dry soil) or *mudflows* (if saturated with water). A kind of wet-soil flow especially characteristic of alpine regions is a result of near-surface melting of frozen soil, while layers below remain frozen and impermeable. The resultant movement of sodden material over solid, impermeable ground is **solifluction.** A wide variety of materials—soil, rocks, trees, and so on—incorporated together in a single flow is called a **debris avalanche** (figure 19.11). Regardless of the nature of the materials moved, all flows have in common the chaotic movement of the particles or objects in the flow.

Scales and Rates of Movement

Mass movements can occur on a variety of scales, as well as at a variety of rates. They may involve a few cubic meters of material or more than a billion cubic meters. Total displacement may be over a distance of only a few centimeters or even millimeters, or the falling, sliding, or flowing material may travel several kilometers.

In the most rapid mass movements, which include most rockfalls, avalanches, and mudflows, materials can travel at speeds up to several tens of meters per second. Because there is little time for anyone to react once these events start, they are associated with the greatest proportion of mass-movement casualties. Slides generally move at more moderate rates, in the range of a few meters per week to meters per day, and slow slides, slumps, and earthflows may move as slowly as a millimeter per year, or less. Extremely slow movement is described as **creep.** Soil creep is more common than rock creep. The principal impact of the slower kinds of mass wasting is property damage.

Recognizing Past Mass Movements

How can past mass movements be recognized? Past rockfalls can be quite obvious, especially in vegetated areas. Talus slopes are inhospitable to most vegetation, so rockfalls tend to remain barren of trees and plants, as in figure 19.8. Lack of vegetation may also mark the paths of past debris avalanches or other soil flows or slides (figure 19.12). These scars on the landscape point plainly to slope instability.

Landslides are not the only kinds of mass movements that recur in the same places. Snow avalanches disappear when the snow melts, but historical records of past avalanches can pinpoint particularly dangerous areas. Records of the character of past volcanic activity and an examination of the typical products of a particular volcano can similarly be used to assess that volcano's tendency to produce nuées ardentes.

Very large slumps and slides may be less obvious, especially when viewed from ground level. The coherent nature of rock and soil movement in most slides means that vegetation growing atop the slide may not be greatly disturbed by the movement. Aerial photography or high-quality topographic maps can be helpful in such a case. In a regional overview, the mass movement often shows up very clearly, revealed by a scarp at the head of the slump or an area of lumpy, disrupted topography relative to surrounding, more stable areas.

With creep, individual movements are short-distance and the whole process is slow, so vegetation may continue to grow in spite of the slippage. More detailed observation, however, can reveal the movement. For example, trees are biochemically "programmed" to grow vertically upward. If the soil in which they are growing begins to creep downslope, the tree trunks may be tilted, and this indicates the soil movement. Further tree growth will continue to be vertical. If slow creep is prolonged over a considerable period of time, during which growth proceeds, curved tree trunks may result (figure 19.13). Inanimate objects can

Figure 19.12 Areas prone to landslides may be recognized by the failure of vegetation to establish itself on unstable slopes. Rock Creek valley, Beartooth Mountains.

Figure 19.13 Curved tree trunks develop on an imperfectly stabilized road cut, Lassen Park, California.

reflect soil creep also. Slanted utility poles and fences and the tilting of once-vertical gravestones or other monuments likewise indicate that the soil is moving. The ground surface itself may show cracks parallel to (across) the slope.

Landslides and Human Activities

Human activities can increase the risk of landslides in many ways. One is by the clearing of stabilizing vegetation. In some instances, where clear-cutting logging operations have exposed sloping soil, for example, earthflows and mudflows have become far more frequent or been more severe than before. Activities such as dam construction may, alternatively, modify pore pressures, and increased pore pressure increases landslide risk, as was seen in the Vaiont Dam disaster, described in box 19.1

Many types of construction lead to oversteepening of slopes. Highway roadcuts, quarrying or open-pit mining operations, and construction of stepped home-building sites on hillsides are among the activities that can cause problems (figure 19.14). Slopes cut in unconsolidated material at angles higher than the angle of repose of that material are by nature unstable. Where dipping layers of rock are present, removal of material at the lower end of the layers may leave large masses of rock unsupported, held in place only by friction between the layers.

Building a house above an unstable slope adds weight to the slope, thereby increasing the shearing stress acting on the slope. In addition, watering the lawn, using a septic tank for sewage disposal, and even putting in an in-ground swimming pool from which water leaks slowly are all activities that increase the soil's moisture content and can render the slope more susceptible to sliding. Irrigation and use of septic systems both increase the flushing of fresh water through soils and sediments. In areas underlain by quick or sensitive clays, these practices may hasten the washing-out of salty pore

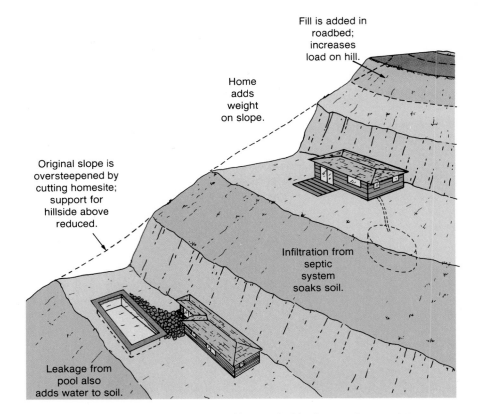

Figure 19.14 Effects of construction and human habitation on slope stability.

waters and the destabilization of the clays. On the other hand, the planting of vegetation associated with deliberate landscaping can reduce the risk of sliding.

Possible Preventive Measures

In places where the structures to be protected are few or small, and the slide zone is narrow, it may be economically feasible to bridge structures and simply let the slides flow over them. For example, this might be done to protect a railway line or road running along a valley from avalanches—either snow or debris—below a particularly steep slope (figure 19.15). This solution would be far too expensive on a large scale, however, and no use at all if the base on which the structure was built were sliding also.

If a slope is too steep to be stable under the load it carries, then one can either (1) reduce the slope angle, (2) place additional supporting material at

the foot of the slope to prevent a slide or flow at the base of the slope, or (3) reduce the load (weight) on the slope by removing some of the rock or soil (or artificial structures) high on the slope. See, for example, figures 19.16 and 19.17. These measures may be used in combination. Where retaining walls are placed below a slope, the greatest success has generally been achieved with low, thick walls placed at the toe of a relatively coherent slope; see, by contrast, figure 19.18.

The other principal strategy for reducing landslide hazards is to decrease the water content, or pore pressure, of the rock or soil, thereby increasing frictional resistance to sliding. This might be done by covering the surface completely with an impermeable material and diverting surface runoff above the slope. Alternatively, subsurface drainage might be undertaken. Moisture-reducing techniques have the greatest impact where rocks or soil are relatively permeable and drain readily.

Anchoring an unstable mass is another approach. Vertical piles driven

CHAPTER 20

Mineral and Energy Resources

Outline

Wind has been an alternative energy source for centuries, but the technology has changed greatly over that time; here, the world's largest vertical-axis wind turbine, Brushland, Texas.
© Wm. C. Brown Communications/Doug Sherman, photographer

Introduction

We are all affected by issues relating to geologic resources—their availability, their cost, the consequences of their use. This chapter offers a very brief introduction to our mineral and energy resources, including a look at what may lie ahead. The chapter begins with a look at the occurrences of a variety of rock and mineral resources and then briefly considers the U.S. and world supply-and-demand picture. Aspects of that picture suggest a need to develop additional sources of mineral materials for the future, and several possibilities are examined. Environmental impacts of mining activities are also noted. We then take a similar look at the energy sources currently in most common use, and finally survey briefly some additional energy sources that may become increasingly important in the future.

Basic Concepts: Reserves and Resources

The term *resources* is often used informally to designate some natural materials that are useful and/or economically valuable to present society. In the context of minerals and fuels, the term has a more precise definition, and is distinguished from the related term *reserves.*

The amount represented by the **reserves** is that quantity of a given material that has been found and could be exploited economically with existing technology. The term is usually further restricted to apply only to material not yet consumed. (The term *cumulative reserves* encompasses the quantity of that material already used, as well as remaining reserves.) The category of **resources** is a much larger one: It includes reserves, plus deposits of the material that are known but that cannot be exploited economically given present prices and technology, plus deposits that have not yet been found but are believed to exist, based on extrapolation from the abundance and geologic occurrence of known deposits. Clearly, the reserves are the most conservative estimate of the amount of a given mineral or fuel available, especially in the near future.

Because economic considerations enter into the definition of reserves, changing economic factors can cause particular mineral or fuel deposits to be reclassified as reserves that were previously classified only as part of resources. For example, increases in demand, or changes in other factors that drive up prices, make some previously uneconomic mineral deposits profitable to mine. Improvements in technology that decrease the cost of extracting the particular material of interest likewise affect the profitability of developing particular deposits.

In the context of energy, it is also important to distinguish between *renewable* and *nonrenewable* energy sources. **Nonrenewable** resources are those that are not being produced at present or that are being produced at rates much slower than current consumption rates. On a human time scale, the supplies of nonrenewable fuels are finite. As these resources become depleted, there is increasing interest in more extensive development of **renewable** resources, those that either can be replenished on a human time scale (such as wood) or those that can be used without actual depletion of supply (like solar energy).

Types of Mineral Deposits

The bulk of the earth's crust is composed of fewer than a dozen elements. In fact, eight chemical elements make up more than 98 percent of the crust. Many of the elements *not* found in abundance in the earth's crust, including industrial and precious metals, essential components of chemical fertilizers, and elements like uranium that serve as energy sources, are vitally important to society. Some of these are found in very minute amounts in the average rock of the continental crust: copper, 0.006 percent; tin, 2 parts per million (ppm); gold, 4 parts per *billion* (ppb). Clearly, then, many useful elements must be mined from very atypical rocks.

What Is an Ore?

An **ore** is a rock in which a valuable or useful metal occurs at a concentration sufficiently high to make it economically worth mining. A given ore deposit may be described in terms of the **enrichment** or **concentration factor** of a particular metal:

$$\text{concentration factor (of a given metal in an ore)} = \frac{\text{concentration of the metal in that ore}}{\text{concentration of that metal in average continental crust}}$$

The higher the concentration factor, the richer the ore (by definition), and the less of it needs to be mined to extract a given amount of metal.

The unit value of the mineral or metal extracted and its concentration in a particular deposit are major factors determining the profitability of mining a specific deposit. Naturally, the economics are sensitive to world demand. If demand climbs and prices rise in response, additional, less-enriched ore deposits may be opened up; a fall in price causes economically marginal mines to close. The practicality of mining a specific ore body may also depend on the mineral(s) in which a metal of interest is found because this affects the cost of extracting the pure metal. Three iron deposits containing equal concentrations of iron are not equally economic if the iron occurs mainly in oxides in one deposit, in silicates in another, and in sulfides in the third.

Because ores are somewhat unusual rocks, it is not surprising that the known economic mineral deposits are very unevenly distributed around the world. (See, for example, figure 20.1.) Note in that figure the highly variable

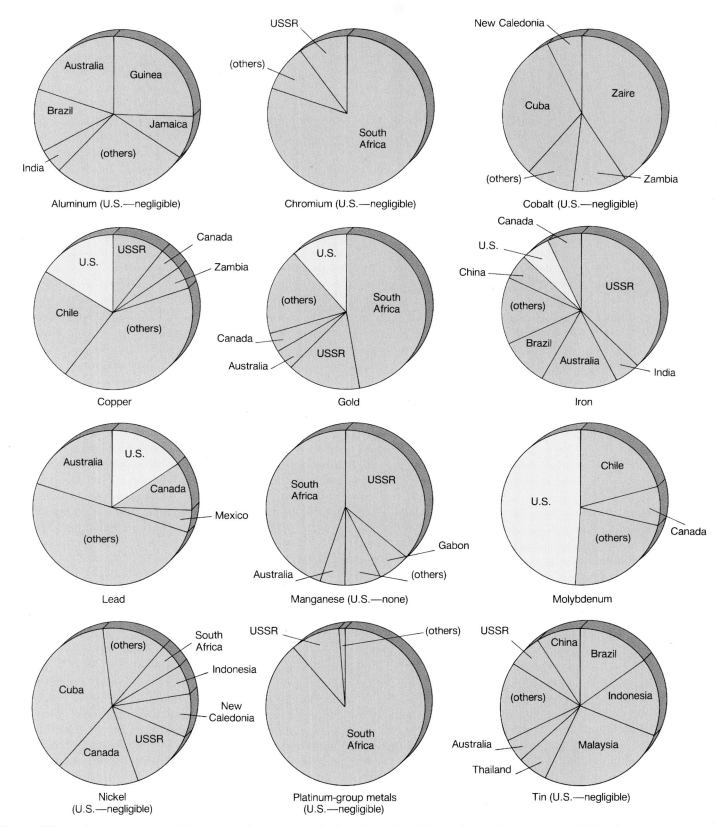

Figure 20.1 Proportions of world reserves of some nonfuel minerals controlled by various major producers. Although the United States is a major consumer of most metals, it is a major producer of very few.

Source: Data from *Mineral Commodity Summaries 1990*, U.S. Bureau of Mines

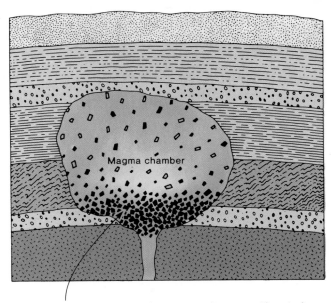

A dense mineral like chromite or magnetite may settle out of a crystallizing magma to be concentrated at the bottom of the chamber.

Figure 20.2 Formation of a magmatic ore deposit by gravitational settling of a dense mineral during crystallization.

Figure 20.3 Effects of hydrothermal activity along the East Pacific Rise. Sample of minerals deposited around a black smoker, including metallic sulfides.
Photograph by W. R. Normark, USGS Photo Library, Denver, Colorado

proportion of different mineral reserves occurring in the United States and elsewhere.

Magmatic Deposits

Magmatic activity gives rise to several kinds of deposits. Certain igneous rocks, just by virtue of their compositions, contain high concentrations of useful silicates or other minerals. These deposits may be especially valuable if the rocks are coarse-grained *pegmatites,* from which individual minerals are easily recovered (recall figure 3.9). In some pegmatites, crystals may be over 10 meters (30 feet) long. Many pegmatites are also enriched in uncommon elements. Pegmatites commonly crystallize from the residual fluids left after most of a body of magma has solidified; many rare or trace elements not readily incorporated into the crystal structures of the common silicates are likewise left over, concentrated in the residual fluids at this stage. Among these are lithium, boron, beryllium, and uranium. Rarer minerals mined from pegmatites include tourmaline and beryl. (Tourmaline is used as a gemstone and for crystals in radio equipment, and is also

mined for the element boron that it contains. Beryl is mined for the metal beryllium when the crystals are of poor quality, or for the gemstones aquamarine and emerald when found as large, clear, well-colored crystals.)

Other useful minerals may be concentrated within a cooling, crystallizing magma chamber by gravity. If they are more or less dense than the magma, they may sink or float as they crystallize, instead of remaining suspended in the solidifying silicate mush, and accumulate in thick layers that are easily mined (figure 20.2). This, then, is a form of fractional crystallization (chapter 3) that produces economic mineral deposits. Chromite (chromium oxide, Cr_2O_3) and magnetite (Fe_3O_4) are both quite dense. In a magma of suitable bulk composition, rich concentrations of these minerals may form in the lower part of the magma chamber by gravitational settling during crystallization.

Hydrothermal Ores

Not all mineral deposits related to igneous activity form within igneous rock bodies. Magmas have water and other fluids dissolved in or associated with

them. Particularly during the later stages of crystallization, the fluids may escape from the cooling magma, seeping through cracks and pores in the surrounding rocks, carrying with them dissolved salts, gases, and metals. These warm fluids can leach additional metals from the rocks through which they pass. In time, the fluids cool and deposit their dissolved minerals, creating a **hydrothermal** (literally, "hot water") ore deposit (figure 20.3).

The particular minerals deposited vary with the composition of the hydrothermal fluids, but a great variety of metals occur in hydrothermal deposits worldwide: copper, lead, zinc, gold, silver, platinum, uranium, and others. Because sulfur is a common constituent of magmatic gases and fluids, the ore minerals are frequently sulfides.

The hydrothermal fluids need not all originate within the magma. Sometimes, circulating subsurface waters are heated sufficiently by a nearby cooling magma to dissolve, concentrate, and redeposit valuable metals in a hydrothermal ore deposit. The fluid involved may also be a mix of magmatic and nonmagmatic fluids.

Table 20.2   World and U.S. production and reserves statistics, 1989*

Material	Production	World reserves	Projected lifetime (years)	U.S. reserves	Projected lifetime (years)
bauxite	111,000	25,300,000	228	41,800	8.7
chromium	13,200	1,133,000	85.8	0	0
cobalt	42	3,641	86.7	0	0
copper	9,710	387,000	39.9	62,700	25.3
iron ore	1,040,000	162,000,000	156	17,700,000	221
lead	3,800	77,000	20.3	12,100	24.4
manganese	26,500	900,000	34.0	0	0
nickel	925	54,000	58.4	37	0.2
tin	231	4,710	20.4	22	0.4
zinc	7,740	162,000	20.9	22,100	68
gold	2,100	46,200	22.0	5,320	37.2
silver	15,400	308,000	20.0	34,100	5.8
platinum group	594	61,600	104	275	3.6
gypsum	105,200	2,600,000	24.7	800,000	29.6
phosphate	187,000	15,240,000	81.5	1,430,000	30.4
potash	34,300	18,700,000	545	99,000	16.5
sulfur	64,400	1,540,000	23.9	158,000	11.3

Source: *Mineral Commodity Summaries 1990*, U.S. Bureau of Mines.

*Reserves in thousands of tons, except for gold, silver, and platinum-group metals, for which figures are in tons.

†Note that bauxite consumption is only a partial measure of total aluminum consumption; additional aluminum is consumed as refined aluminum metal, of which there are no reserves.

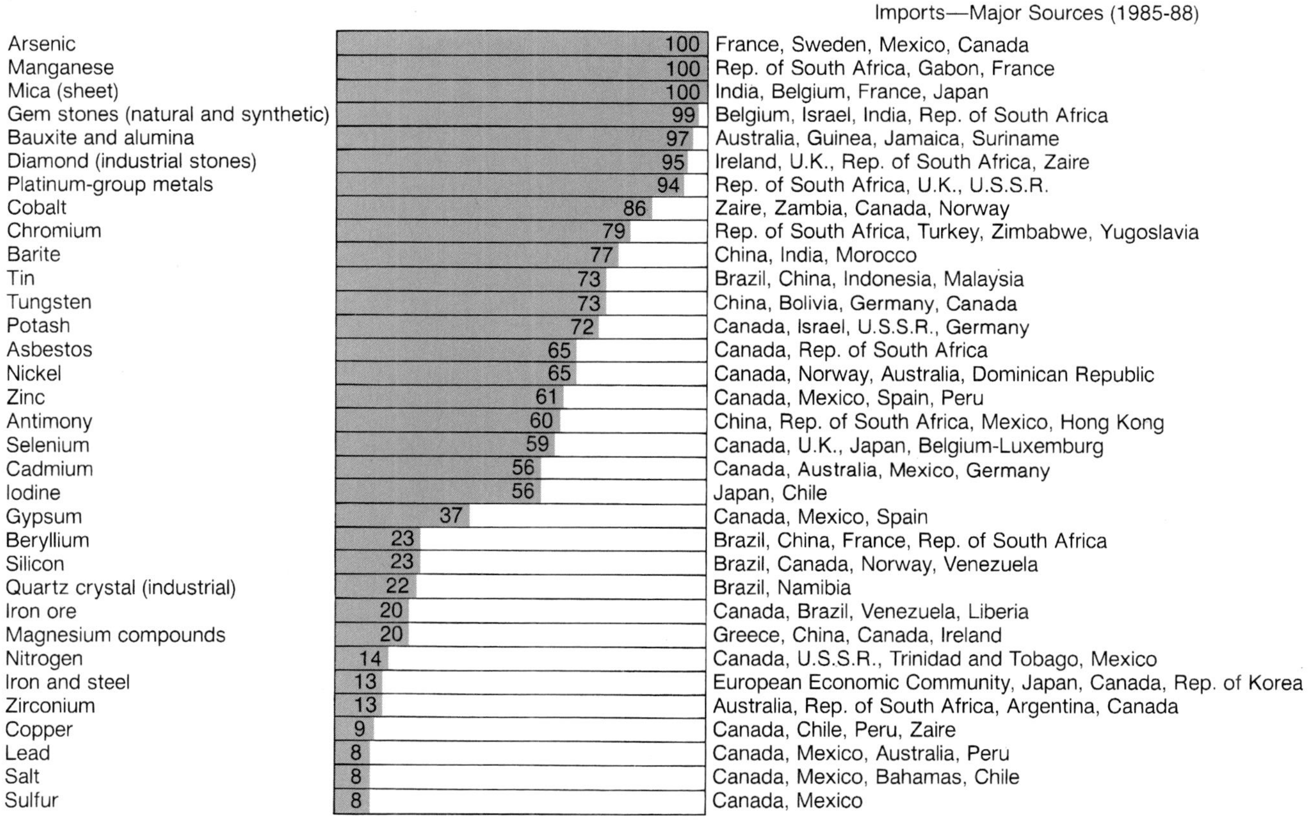

Imports—Major Sources (1985-88)

Material	%	Major Sources
Arsenic	100	France, Sweden, Mexico, Canada
Manganese	100	Rep. of South Africa, Gabon, France
Mica (sheet)	100	India, Belgium, France, Japan
Gem stones (natural and synthetic)	99	Belgium, Israel, India, Rep. of South Africa
Bauxite and alumina	97	Australia, Guinea, Jamaica, Suriname
Diamond (industrial stones)	95	Ireland, U.K., Rep. of South Africa, Zaire
Platinum-group metals	94	Rep. of South Africa, U.K., U.S.S.R.
Cobalt	86	Zaire, Zambia, Canada, Norway
Chromium	79	Rep. of South Africa, Turkey, Zimbabwe, Yugoslavia
Barite	77	China, India, Morocco
Tin	73	Brazil, China, Indonesia, Malaysia
Tungsten	73	China, Bolivia, Germany, Canada
Potash	72	Canada, Israel, U.S.S.R., Germany
Asbestos	65	Canada, Rep. of South Africa
Nickel	65	Canada, Norway, Australia, Dominican Republic
Zinc	61	Canada, Mexico, Spain, Peru
Antimony	60	China, Rep. of South Africa, Mexico, Hong Kong
Selenium	59	Canada, U.K., Japan, Belgium-Luxemburg
Cadmium	56	Canada, Australia, Mexico, Germany
Iodine	56	Japan, Chile
Gypsum	37	Canada, Mexico, Spain
Beryllium	23	Brazil, China, France, Rep. of South Africa
Silicon	23	Brazil, Canada, Norway, Venezuela
Quartz crystal (industrial)	22	Brazil, Namibia
Iron ore	20	Canada, Brazil, Venezuela, Liberia
Magnesium compounds	20	Greece, China, Canada, Ireland
Nitrogen	14	Canada, U.S.S.R., Trinidad and Tobago, Mexico
Iron and steel	13	European Economic Community, Japan, Canada, Rep. of Korea
Zirconium	13	Australia, Rep. of South Africa, Argentina, Canada
Copper	9	Canada, Chile, Peru, Zaire
Lead	8	Canada, Mexico, Australia, Peru
Salt	8	Canada, Mexico, Bahamas, Chile
Sulfur	8	Canada, Mexico

Figure 20.8   Proportions of U.S. mineral needs supplied by domestic sources and by imports, as a percentage of apparent consumption.

Source: Data from *Mineral Commodity Summ.*

New Methods in Mineral Exploration

Most of the "easy ores," near-surface mineral deposits in readily accessible places, have probably already been discovered. Fortunately, however, a variety of new methods are being applied in the search for the less easily located ore deposits. Geophysics, for example, provides some assistance. Rocks and minerals vary in density and in electrical and magnetic properties, so changes in rock types or distribution below the earth's surface can cause gravitational and magnetic anomalies that can be measured at the surface, as well as small variations in the electrical conductivity of rocks. Some ore deposits may be detected in this manner. As a simple example, because many iron deposits are strongly magnetic, large magnetic anomalies may indicate the presence and the extent of a subsurface body of iron ore. Radioactivity is another readily detected property; uranium deposits can be located with the aid of a Geiger counter or scintillation counter.

Geochemical prospecting adds other methods that are increasingly widely used. Some studies are based on the recognition that soils reflect, to a degree, the chemistry of the materials from which they formed. The soil over a copper-rich ore body, for instance, may itself be enriched in copper relative to surrounding soils. Occasionally, plants can be sampled instead of soil. Certain plants tend to concentrate particular metals, making the plants very sensitive indicators of locally high concentrations of those metals. (For example, the "locoweed" that causes strange symptoms in grazing cattle does so because it concentrates the toxic element selenium.) Water or stream sediments may likewise be sampled and analyzed for high concentrations of the metals being sought. Even soil gases can supply clues to ore deposits. Mercury is a very volatile metal, and high concentrations of mercury vapor have been found in gases filling soil pore spaces over mercury ore bodies.

Remote sensing, which includes the use of aerial photographs and satellite imagery, is another useful tool for mineral exploration. Landsat (now Eosat) satellites provide images covering the world every eighteen days. This gives at least a preliminary look at many areas that might otherwise be too remote or difficult to reach on the ground. Moreover, the satellite images can be processed through computers to sharpen the images or to focus on unusual features in the geology or vegetation that might bear further investigation (figure 20.9). Remote sensing must still be followed by ground-based geologic studies eventually. Perhaps its greatest usefulness at present is in limiting the scope of these slower and more costly detailed studies, by pinpointing areas particularly likely to yield new mineral deposits and thus allowing further exploration efforts to be concentrated in the most promising areas.

Advances in geologic understanding also play a role in mineral exploration. Development of plate-tectonic theory has helped geologists to recognize the association between particular types of plate boundaries and the occurrence of certain kinds of mineral deposits. This association may direct the search for new ores. For example, because geologists know that molybdenum deposits are often found over existing subduction zones, they can logically explore for more molybdenum deposits in other present or past subduction zones. Also, the realization that many of the continents were once united suggests likely areas to look for more ores (figure 20.10). If mineral deposits of a particular kind are known to occur near the margin of one continent, similar deposits might be found at the corresponding edge of another continent that was once connected to the first. If a mountain belt on one continent is rich in some kind of ore and seems to have a counterpart range on another continent that was once linked to the first, the same kind of ore may occur in the matching mountain belt. Such reasoning is partly responsible for the supposition that economically valuable ore deposits may exist on Antarctica. The most successful exploration efforts may integrate several of the methods just described, making use of the particular advantages of each.

Marine Mineral Resources

Virtually every chemical element is dissolved in seawater. However, most of the dissolved material is halite. Vast volumes of seawater would have to be processed to extract small quantities of such metals as copper and gold. Moreover, current technology is not adequate to extract, selectively and efficiently, a few specific metals of interest from all that salt water on a routine basis, although the Japanese have been experimenting with extraction of uranium from seawater. Other types of underwater mineral deposits have greater potential.

As noted in chapter 17 on glaciers, during the last ice age, when a great deal of water was locked in ice sheets, eustatic sea levels were lower than at present. Much of the now-submerged continental shelf area was dry land, and streams running off the continents flowed out across the exposed shelves to reach the sea. Where the streams' drainage basins included appropriate source rocks, those streams might have concentrated valuable placer deposits. With the melting of the ice and rising sea levels, any placers on the continental shelves were submerged. Finding and mining these deposits may be more costly and difficult than land-based mining, but as land deposits are exhausted, placer deposits on the continental shelves may well be worth seeking.

The hydrothermal ore deposits forming along some seafloor spreading ridges are another possible source of needed metals. In many places, the quantity of material being deposited is too small, and the depth of water above would make recovery prohibitively expensive, at least over the near term. However, the metal-rich muds of the Red Sea contain sufficient concentrations of such metals as copper, lead, and zinc that some exploratory dredging is underway, and several companies are interested in the possibility of mining those sediments. Along a section of the Juan de Fuca Ridge, off the west coast of Oregon and Washington, hundreds of thousands of tons of zinc- and silver-rich sulfides have already been deposited, and the hydrothermal activity continues.

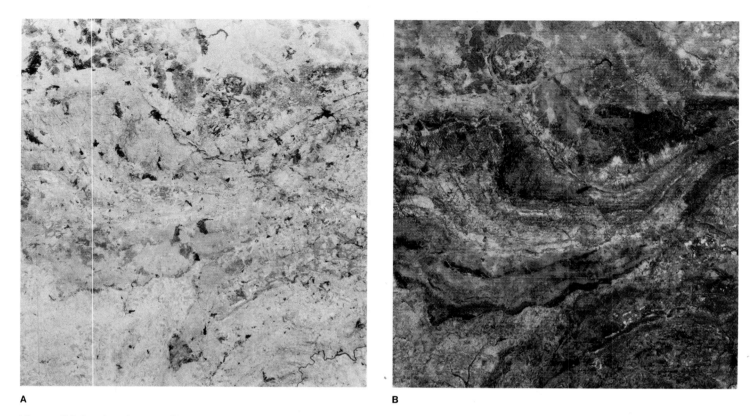

A   **B**

Figure 20.9   Landsat satellite photographs may reveal details of the geology that will aid in mineral exploration. Vegetation, also sensitive to geology, may enhance the image. (*A*) View of South Africa, dry season. (*B*) Same view, rainy season. Recognizable geologic features include a granite pluton (round feature at top center) and folded layers of sedimentary rock (below).
(*A*) and (*B*) © NASA

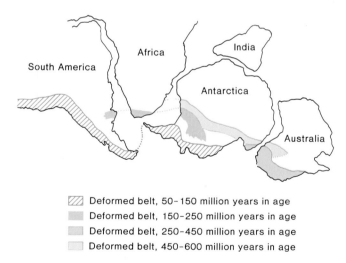

Deformed belt, 50–150 million years in age
Deformed belt, 150–250 million years in age
Deformed belt, 250–450 million years in age
Deformed belt, 450–600 million years in age

Figure 20.10   Pre-continental-drift reassembly of landmasses suggests locations of possible ore deposits in unexplored regions by extrapolation from known deposits on other continents.

Adapted with permission from C. Craddock, et al., Geologic Maps of Antarctica. *Antarctic Map Folio Series,* Folio 12. Copyright © 1970 by the American Geographical Society, New York, NY.

Perhaps the most widespread undersea mineral resources, and the most frequently and seriously discussed as near-term resources, are the *manganese nodules.* Ranging in size up to about 10 centimeters in diameter, these are lumps composed mostly of manganese oxides and hydroxides. They also contain lesser amounts of iron, copper, nickel, zinc, cobalt, and other metals. Indeed, the value of the minor metals may be a greater financial motive for mining manganese nodules than the manganese itself. The nodules are found over much of the deep-sea floor, in regions where sedimentation rates are slow enough not to bury them (recall figure 13.17). At present, the costs of recovering these nodules are high compared with the costs of mining the same metals on land, and the technical problems associated with recovering the nodules from beneath several kilometers of water remain to be worked out, as does the practical question of who owns seabed resources in international waters and who has the right to exploit them.

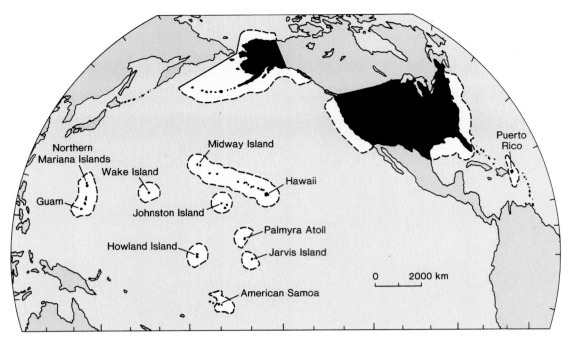

Figure 20.11 The Exclusive Economic Zone of the United States.
Source: *U.S. Geological Survey 1983 Annual Report.*

The 1982 U.N. Law of the Sea Treaty established **Exclusive Economic Zones** (EEZs) extending up to 200 miles from the shorelines of coastal nations, within which those nations have exclusive rights to mineral-resource exploitation. These areas are not restricted to continental shelves only. Some deep-sea manganese nodules fall within Mexico's EEZ; Saudi Arabia and the Sudan have the rights to the metal-rich muds on the floor of the Red Sea. Included in the EEZ off the west coast of the United States is a part of the East Pacific Rise spreading-ridge system, with its active hydrothermal vents and sulfide deposits. Areas of the ocean basins outside any nation's EEZ are under the jurisdiction of an International Seabed Authority, and the treaty includes some provision for financial and technological assistance to developing countries wishing to share in resources there.

The United States did not sign the treaty, but in March 1983, President Reagan unilaterally proclaimed a U.S. Exclusive Economic Zone that extends to the 200-mile limit offshore. This move expands the undersea territory under U.S. jurisdiction to some 3.9 billion acres, more than 1½ times the land area of the United States and its territories (figure 20.11). So far, however, exploration of this new domain has been minimal.

Table 20.3	Metal Recycling in the United States, 1985–89 (recycled scrap as percentage of consumption)				
Metal	**1985**	**1986**	**1987**	**1988**	**1989**
aluminum	16.4	15.2	15.6	19.5	21.6
chromium	24.9	20.7	24.0	22.5	20.9
cobalt	5.7	15.2	14.1	14.0	14.5
copper	23.5	22.3	22.7	23.4	24.9
lead	50.3	50.1	54.7	56.4	60.3
manganese	0	0	0	0	0
nickel	27.2	25.3	24.6	32.3	34.4
platinum group	38.6	42.7	54.2	52.1	72.1
zinc	5.5	7.0	8.1	8.8	9.5

Source: *Mineral Commodity Summaries 1990*, U.S. Bureau of Mines.

Recycling

The most effective way to extend some mineral reserves may be through recycling, which reduces the need for additional primary production from reserves. Some metals are already extensively recycled, at least in the United States (see table 20.3). Worldwide, recycling is less widely practiced, in part because the less-industrialized countries have accumulated fewer manufactured products from which materials can be retrieved. Two additional benefits of recycling are a reduction in the volume of waste-disposal problems and a decrease in the extent to which more land must be disturbed by new mining activities.

Unfortunately, not all materials lend themselves equally well to recycling. Among those that work out best are metals that are used in pure form in large chunks—copper in pipes and wiring, lead in batteries, and aluminum in beverage cans. The individual metals are relatively easy to recover and require minimal effort to purify for reuse. Recycling aluminum is also appealing from an energy (and cost) standpoint: It takes only one-twentieth as much

Figure 20.12 Collapse of land surface over an old abandoned copper mine in Arizona. For scale, note roads, trees, and houses.

Photograph by F. L. Ransome, courtesy of USGS Photo Library, Denver, Colorado

Figure 20.13 The world's largest open-pit mine, Bingham Canyon, Utah. Over $6 billion worth of minerals has been extracted from this one mine over the decades of its operation.

Photograph courtesy of Kennecott

energy to produce aluminum by recycling old scrap as it does to extract aluminum metal from bauxite. Sometimes, it is even economical to recover and recycle some of these discrete metallic items from old dump or landfill sites.

Where different materials are intermingled in complex manufactured objects, it is more difficult and costly to extract individual metals. Consider trying to separate the various metals of a refrigerator, a lawn mower, or a television set. Only in a few rare cases are metals valuable enough that the recycling effort in such a case may be worthwhile; for example, there is in-terest in recovering the platinum from catalytic mufflers. Alloys, mixtures of different metals in application-specific proportions, present additional problems.

Also, some materials are not used in discrete objects at all. The potash and phosphorous in fertilizers are strewn across the land and cannot be recovered. Road salt washes off of streets and into soil and storm sewers. Clearly, these things cannot be recovered and reused. For the foregoing and other reasons, it is unrealistic to expect that all mineral materials can ever be completely recycled.

Impacts of Mining and Mineral-Processing Activities

Mining and mineral-processing activities can modify the environment in a variety of ways. Most obvious is the presence of the mine itself. Both underground mines and surface mines have their own sets of associated impacts.

Underground mines are relatively inconspicuous. They disturb a small land area close to the principal shaft(s). Waste rock dug out of the mine may be piled close to the mine's entrance, but in most underground mines, the tunnels follow the ore body as closely as possible to minimize the amount of non-ore rock to be removed. When mining activities are complete, the shafts can be sealed, and the area often returns very nearly to its pre-mining condition. However, near-surface underground mines occasionally have collapsed years after abandonment, when supporting timbers have rotted away or groundwater has enlarged the underground cavities through solution (figure 20.12). In some cases, the collapse occurred so long after the mining had ended that the existence of the old mines had been entirely forgotten.

Surface-mining activities consist of either open-pit mining (including quarrying) or strip-mining. Open-pit mining, such as is shown in figure 20.13, is practical when a large three-dimensional ore body is located near the surface. Most of the material in the

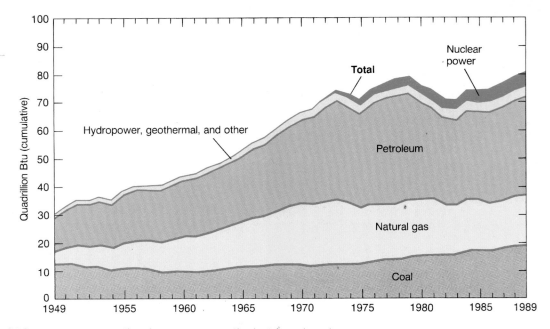

Figure 20.14 U.S. energy consumption, by source, over the last four decades.

Sources: Data from *Annual Energy Review 1987* and *Monthly Energy Review, December 1989,* U.S. Energy Information Administration, Department of Energy.

pit is the valuable commodity and is extracted for processing. Thus, this procedure permanently changes the topography, leaving a large hole in its wake. The exposed rock may begin to weather and, depending on the nature of the ore body, may release pollutants into surface runoff water.

Strip mining, more often used to extract coal than mineral resources, is practiced most commonly when the material of interest occurs in a layer near and approximately parallel to the surface. Overlying vegetation, soil, and rock are stripped off, the coal or other material is removed, and the waste rock and soil cover is dumped back as a series of **spoil banks.** The finely broken-up material of spoil banks, with its high surface area, is very susceptible to both erosion and chemical weathering. Strip-mining is discussed further later in this chapter, in the context of coal.

The **tailings,** piles of crushed and ground rock that are left over from mineral processing, may end up heaped around the processing plant to weather and wash away, much like spoil banks. As noted earlier, traces of the ore are left behind in tailings. Depending on the nature of the ore, rapid weathering of the tailings may leach out such

harmful elements as mercury, arsenic, cadmium, and uranium, contaminating surface and groundwaters.

The chemicals used in processing—for example, the cyanide used on gold ore—are often toxic also, and there is concern about possible hazards associated with large heap-leaching operations. Smelting to extract metals from ores may release not only sulfurous gases from sulfide ores, but also lead, arsenic, mercury, and other potentially toxic, volatile elements along with exhaust gases and ash, unless emissions are tightly controlled. Mineral processing, then, represents a huge potential source of pollutants, if not done with care.

Our Primary Energy Sources: Fossil Fuels

The term *fossil* refers to any remains or evidence of ancient life. The **fossil fuels,** then, are those energy sources that formed from the remains of once-living organisms. These include oil, natural gas, coal, and fuels derived from oil shale and tar sand. Oil and natural gas are currently our two principal energy sources (figure 20.14). The differences in the physical properties

of the various fossil fuels arise from differences in the starting materials from which they formed and in what happened to those materials after the organisms died and were buried in the earth.

Oil and Natural Gas

Oil, or petroleum, is not a single chemical compound. Petroleum comprises a variety of liquid hydrocarbon compounds (compounds made up of various proportions of the elements carbon and hydrogen). There are also gaseous hydrocarbons, (so-called natural gas), of which the compound methane (CH_4) is the most common. How organic matter is transformed into liquid and gaseous hydrocarbons is not fully understood, but the transformation is believed to occur somewhat as described below.

Microscopic life is abundant over much of the oceans. When these organisms die, their remains settle to the sea floor. Near shore—for example, on many continental shelves—terrigenous sediments accumulate rapidly. In such a setting, the starting requirements for the formation of oil are satisfied: An abundance of organic matter is rapidly buried by sediment and so is

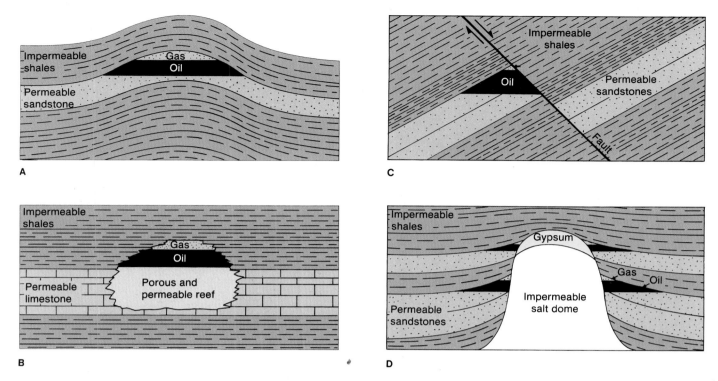

Figure 20.15 Types of petroleum traps. (*A*) A simple fold trap. (*B*) Petroleum accumulated in a fossilized ancient coral reef. (*C*) A fault trap. (*D*) Petroleum trapped against an impermeable salt dome that has risen up from a buried evaporite deposit.

prevented from being destroyed by reaction with oxygen. Oil and most natural gas deposits are believed to form from such accumulations of marine microorganisms. (Additional natural gas deposits may be derived from the remains of terrestrial plants.)

As burial continues, the organic matter begins to change. Pressures increase as the weight of the overlying pile of sediment or rock increases; temperatures increase according to the geothermal gradient; and slowly, over long periods of time, chemical reactions occur. These reactions break down the large, complex organic molecules into simpler, smaller hydrocarbon molecules. The nature of the hydrocarbons changes with time and with continued heat and pressure. In the early stages of petroleum formation, the deposit may consist mainly of larger hydrocarbon molecules ("heavy" hydrocarbons), which have the thick, nearly solid consistency of asphalt. As the petroleum matures, and the breakdown of large molecules continues, successively "lighter" hydrocarbons are produced. Thick liquids give way to thinner ones, from which are derived

lubricating oils, heating oils, and gasoline. In the final stages, most or all of the petroleum is further broken down into very simple, light, gaseous molecules—natural gas. Most of the maturation process occurs in the temperature range of 50° to 100° C (approximately 120° to 210° F.). Above these temperatures, the remaining hydrocarbon is almost wholly methane; with further temperature increases, methane can be broken down and destroyed in turn.

The amount of time required for oil and gas to form is not known precisely. Since virtually no petroleum is found in rocks younger than 1 to 2 million years old, geologists infer that the process is comparatively slow. Even if it took only a few tens of thousands of years (a geologically short period), the world's oil and gas are being used up far faster than significant new supplies could be produced.

Once the solid organic matter is converted to liquids and gases, the hydrocarbons can migrate out of the rocks in which they formed, if the surrounding rocks are permeable. The pores and cracks in rocks are com-

monly saturated with water at depth. Most oils and all natural gases are less dense than water, so they tend to rise as well as to migrate laterally through the water-filled pores of permeable rocks.

Commercially, the most valuable deposits are those in which a large quantity of oil and/or gas has been concentrated and confined by impermeable rocks (commonly shales) in **traps** (figure 20.15). The **reservoir rocks** in which the oil or gas has accumulated should be relatively porous if a large quantity of petroleum is to be found in a small volume of rock, and should also be relatively permeable so that the oil or gas flows out readily once a well is drilled into the reservoir. Sandstones and limestones are common reservoir rocks. If the reservoir rocks are not naturally very permeable, it may be possible to fracture them artificially with explosives or high-pressure fluids to increase the rate at which oil or natural gas flows through them.

Oil is commonly quantified in terms of units of *barrels* (1 barrel = 42 gallons). Worldwide, estimated remaining

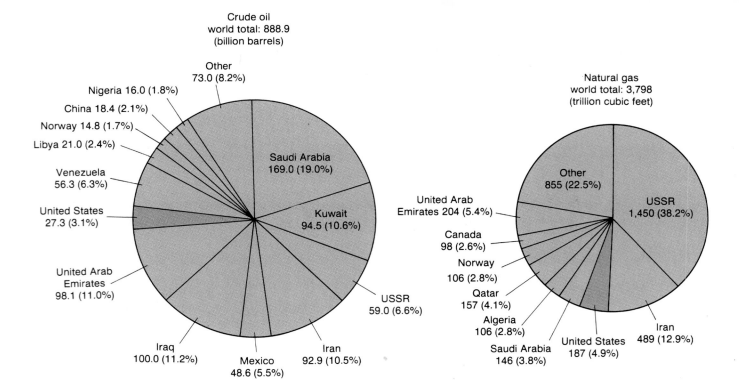

Crude oil
world total: 888.9
(billion barrels)

Other
73.0 (8.2%)
Nigeria 16.0 (1.8%)
China 18.4 (2.1%)
Norway 14.8 (1.7%)
Libya 21.0 (2.4%)
Venezuela
56.3 (6.3%)
United States
27.3 (3.1%)
United Arab
Emirates
98.1 (11.0%)
Saudi Arabia
169.0 (19.0%)
Kuwait
94.5 (10.6%)
USSR
59.0 (6.6%)
Iraq
100.0 (11.2%)
Mexico
48.6 (5.5%)
Iran
92.9 (10.5%)

Natural gas
world total: 3,798
(trillion cubic feet)

United Arab
Emirates 204 (5.4%)
Canada
98 (2.6%)
Norway
106 (2.8%)
Qatar
157 (4.1%)
Algeria
106 (2.8%)
Saudi Arabia
146 (3.8%)
United States
187 (4.9%)
Other
855 (22.5%)
USSR
1,450 (38.2%)
Iran
489 (12.9%)

Note: Quantities are scaled in proportion to area according to the Btu content of the reserves.
One billion barrels of crude oil equals approximately 5.3 trillion cubic feet of wet natural gas.

Figure 20.16 Estimated proven world reserves of crude oil and natural gas, December 1989.
Source: U.S. Energy Information Administration, 1990.

reserves are close to 900 billion barrels (figure 20.16). The United States alone consumes nearly 30 percent of the oil used worldwide. Initially, U.S. oil resources probably amounted to about 10 percent of the world's total. Cumulative U.S. oil resources are estimated to have been not much above 200 billion barrels. Already close to half of that has been produced and consumed; remaining U.S. *resources* are estimated at about 120 billion barrels. Out of that, present proven *reserves* are under 30 billion barrels. For more than a decade, the United States has been using up somewhat more domestic oil each year than the amount of new domestic reserves discovered, or proven, so that net U.S. oil reserves have been decreasing year by year (see table 20.4). Domestic production has likewise been declining. Furthermore, without oil imports, U.S. reserves would dwindle still more quickly. The United States consumes nearly 6 billion barrels of oil per year, and over 40 percent of that amount is imported.

The production/consumption picture for conventional natural gas is similar to that for oil. Natural gas presently supplies nearly 25 percent of the energy used in the United States. The United States has proven natural gas reserves of nearly 200 trillion cubic feet. However, roughly 20 trillion cubic feet are consumed per year, and each year, less is found in new domestic reserves than the quantity consumed. As with oil, the U.S. supply of conventional natural gas is expected to be exhausted within decades, although the gas supplies are less severely depleted than the oil.

While decreasing reserves have prompted exploration of more areas, most regions not yet explored have been neglected precisely because they are unlikely to yield appreciable amounts of petroleum. The high temperatures involved in the formation of igneous and most metamorphic rocks would destroy organic matter, so oil would not have formed or been preserved in these rocks. Nor do these rock

Table 20.4	Proven U.S. reserves of crude oil and natural gas, 1978–87	
Year	Crude oil (billions of barrels)	Natural gas (trillions of cu. ft.)
1978	31.4	208.0
1979	29.8	201.0
1980	29.8	199.0
1981	29.4	201.7
1982	27.9	201.5
1983	27.7	200.2
1984	28.4	197.5
1985	28.4	193.4
1986	26.9	191.6
1987	27.3	187.2

Source: Data from *Annual Energy Review 1988,* U.S. Energy Information Administration, Department of Energy.

types tend to be very porous or permeable, so they generally make poor reservoir rocks as well, unless fractured. The large cratonic regions underlain predominantly by igneous and metamorphic rocks are simply very unpromising places to look for oil.

Despite a quadrupling in oil prices between 1970 and 1980 (after adjustment for inflation), U.S. proven reserves continued to decline. This is further evidence that higher prices do not automatically lead to proportionate, or even significant, increases in fuel supplies. And each time a temporary excess of production over demand causes petroleum prices to plummet, as in early 1986, exploration (as well as the development of new energy sources) comes to a virtual standstill. There is currently limited interest in developing economically uncompetitive oil and gas resources, which include those that can only be extracted using *enhanced-recovery* methods (such as fracturing rocks to increase their permeability or heating them to decrease the oil's viscosity and make it easier to extract), or those technologically difficult to reach (such as gas dissolved in pore waters deep in the crust, under very high pressure).

Coal

Coal is formed, not from marine organisms, but from the remains of land plants. A swampy setting, in which plant growth is abundant and where fallen trees, dead leaves, and other debris are protected from decay (either by standing water or by rapid burial under later layers of plant debris), is especially favorable to the initial stages of coal formation. The Carboniferous period of geologic time, named for its abundant coal deposits, was apparently characterized over much of the world by just the sort of tropical climate conducive to the growth of lush, swampy forests in lowland areas.

Given a suitable setting, the first combustible product formed is *peat*. Peat can form at the earth's surface, and there are places, like Dismal Swamp in North Carolina or the Malaysian peninsula, where peat can be seen forming today. Peat can serve as an energy source—it is so used in Ireland—but it is not a very efficient one. Further burial, with more heat, pressure, and time, gradually dehydrates

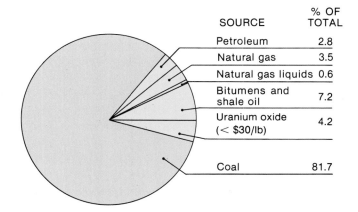

SOURCE	% OF TOTAL
Petroleum	2.8
Natural gas	3.5
Natural gas liquids	0.6
Bitumens and shale oil	7.2
Uranium oxide (< $30/lb)	4.2
Coal	81.7

Figure 20.17 Comparison of the total recoverable energy from various U.S. fossil-fuel reserves. The key word is *recoverable,* which reflects the limitations of present technology.
Source: *The Coal Data Book*, 1980, The President's Commission on Coal.

the organic matter and transforms the spongy peat into soft brown coal *(lignite)* and then into the harder coals *(bituminous* and *anthracite)*. As the coals become harder, their carbon content increases, and so does the amount of heat released by burning a given weight of coal. As with oil, however, the heat to which coals can be subjected is limited: Excessively high temperatures lead to metamorphism of coal into graphite.

Like oil, the higher-grade coals seem to require formation periods that are long compared to the rate at which coal is being used. Consequently, coal, too, can be regarded as a nonrenewable resource. However, the world supply of coal, and that in the United States also, represents a total energy resource far larger than that of petroleum (figure 20.17).

Coal Reserves and Resources

Coal resource estimates are less uncertain than are corresponding estimates for oil and gas. Coal is a solid, so it does not migrate. It is therefore found in the sedimentary rocks in which it formed; one need not seek it in igneous or metamorphic rocks. It occurs in well-defined beds that are easier to map than underground oil and gas concentrations. And because it formed from land plants, which did not become widespread until several hundred million years ago, one need not look for coal in more ancient rocks.

The estimated world reserve of coal is about 650 billion tons; total resources are estimated at over 10 trillion tons. The United States controls over 30 percent of the world reserves, some 200 billion tons. Total U.S. coal resources may approach ten times the reserves. Furthermore, most of that coal is yet unmined. At present, coal provides about 20 percent of the energy used in the United States. While the United States has consumed close to 50 percent of its petroleum resources, it has used up only a few percent of its coal. Even if only the *reserves* are counted, the U.S. coal supply could satisfy U.S. energy needs for several centuries, at current levels of energy consumption, if coal could be used for all purposes. As a supplement to other energy sources, coal could last correspondingly longer.

Coal can also be converted to liquid or gaseous hydrocarbon fuels by causing the coal to react with steam or with hydrogen gas at high temperatures. The conversion processes are termed **liquefaction** (when the product is a liquid fuel) and **gasification** (when the product is gaseous). Both processes are intended to transform the coal into a cleaner-burning, more versatile fuel, thereby expanding its range of possible applications. At present, however, these coal-derived fuels are economically uncompetitive with oil and natural gas.

Oil Shale and Tar Sand

Oil shale is very poorly named. The rock, while always sedimentary, need not be a shale, and the hydrocarbon in it is not oil! The potential fuel in oil shale is a waxy solid, *kerogen,* which is formed from the remains of plants, algae, and bacteria. The rock must be crushed and heated to over 500° C to distill out the "shale oil," which is then refined, somewhat as crude oil is, to produce various liquid petroleum products. The United States has about two-thirds of the world's known supply of oil shale. The total estimated resource could yield 2 to 5 trillion barrels of shale oil. For a number of reasons, we are not yet using this apparently vast resource to any significant extent, and may not be able to do so in the near future.

One reason for this is that much of the kerogen is so widely dispersed through the oil shale that huge volumes of rock must be processed to obtain moderate amounts of shale oil. Even the richest oil shale yields only about three barrels of shale oil per ton of rock processed. The cost of extraction is not presently competitive with that of conventional petroleum. Another problem is that a large part of the oil shale is located at or near the surface and logically would be exploited by strip-mining. Most of the richest oil shale, however, is located in dry areas of the western United States (Colorado, Wyoming, Utah; see figure 20.18), so land reclamation after mining is correspondingly difficult. The water shortage presents a further problem. Current processing technologies require large amounts of water—about three barrels of water per barrel of shale oil extracted. In the western states, there is no obvious abundant source of water to process the oil shale. Finally, since the volume of the rock actually increases during processing, it is possible to end up with a 20 to 30 percent larger volume of waste rock than the volume of rock mined. Aside from problems caused by accelerated weathering of the crushed material, there remains the basic question of where to *put* it. Clearly, it will not all fit back into the space mined out.

Figure 20.18 Outcrops of the Green River Shale, the richest oil shale in the U.S., in Utah. Note the dryness of the area, as shown by sparse native vegetation.
Photograph by R. L. Elderkin, Jr., USGS Photo Library, Denver, Colorado

Tar sands are sedimentary rocks containing a very thick, semisolid, tar-like petroleum. The heavy petroleum in tar sands is believed to have been formed in the same way and from the same materials as lighter oils. Tar-sand deposits may represent very immature petroleum deposits, in which the breakdown of large molecules has not progressed to the production of the lighter liquid and gaseous hydrocarbons. Alternatively, and perhaps more likely, the lighter compounds may simply have migrated away, leaving this dense, viscous material behind. Either way, the tarry petroleum is too thick to flow out of the rock. Like oil shale, tar sand presently must be mined, crushed, and heated to extract the petroleum, which can then be refined into various fuels.

Many of the environmental problems associated with oil shale likewise apply to tar sand. The tar is disseminated through the rock, so large volumes of rock must be mined and processed to extract appreciable petroleum. Many tar sands are near-surface deposits, so the mining method is commonly strip mining. The processing requires a great deal of water, and the amount of waste rock after processing may be larger than the original volume of tar sand.

The United States has very limited tar-sand resources, so it cannot look to tar sand to solve its domestic energy problems, even if the environmental and developmental difficulties could be overcome. The Athabasca tar sands of Canada, however, contain enormous quantities of petroleum, and indeed the bulk of Canadian oil resources are in these tar sands (figure 20.19).

Environmental Impacts of Fossil-Fuel Use

The primary negative impact of petroleum extraction and use relates to oil spills. Spills occur in two principal ways: from accidents during drilling of off-shore oil wells, and from wrecks of oil tankers at sea.

In March 1989, the *Exxon Valdez* ran aground on a reef in Prince William Sound near the Alaskan pipeline port of Valdez. The resultant spill of an estimated 10.2 million gallons of oil left thousands of dead or oil-slicked marine birds and mammals in its wake, caused cancellation of the annual herring season in Valdez, and threatened the salmon fishing as well. The damage would have been far worse had the whole cargo leaked.

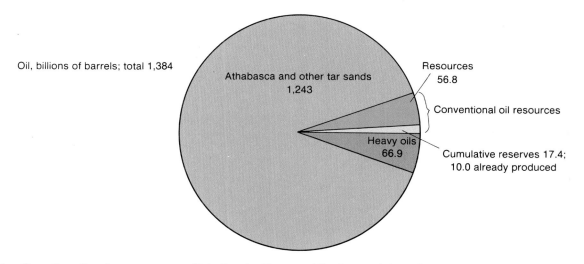

Figure 20.19 Canadian oil and gas resources. Note the significance of the tar-sand deposits.

From R. M. Procter, et al. *Oil and Natural Gas Resources of Canada.* Copyright 1984 Energy, Mines and Resources Canada. Reproduced with the permission of the Minister of Supply and Services Canada, 1991.

When an oil spill occurs, the oil, being less dense than water, floats. The lightest, most volatile hydrocarbons start to evaporate immediately, decreasing the volume of the spill somewhat (and polluting the air). Then a slow decomposition process sets in, due to sunlight and bacterial action. After several months, the spill may be reduced to about 15 percent of its original volume, and what is left is usually thick asphalt lumps. These can persist for years. In calm seas, if a spill is small, it may be contained by floating barriers and picked up by specially designed ships that can skim up to fifty barrels of oil per hour off the water surface. This is clearly a slow process: Three weeks after the *Exxon Valdez* spill, only half a million gallons of oil had been recovered from the sea, and weather conditions threatened to breach the containing barriers at any time; huge quantities of oil had already washed up on and polluted the shore. Some attempts have been made to soak up oil spills with peat moss, wood shavings, and even chicken feathers. The larger spills or spills in rough seas are a particular problem. Perhaps the best prospect for dealing with future oil spills is the development of "oil-hungry" microorganisms that will eat the spill for food and thus get rid of it. Scientists are currently developing suitable bacterial strains.

A major problem posed by coal is the pollution associated with its mining and use. Like all fossil fuels, coal produces carbon dioxide (CO_2) when burned. The additional pollutant that is of special concern with coal is sulfur. The sulfur content of coal can be over 3 percent, some in the form of pyrite (FeS_2), some bound in the organic matter of the coal itself. When the sulfur is burned along with the coal, sulfur gases—notably sulfur dioxide (SO_2)—are produced. These gases are toxic and can be severely irritating to lungs and eyes. The gases also react with water in the atmosphere to produce sulfuric acid, which is removed from the air as acid rain. Some of the sulfur in coal can be removed prior to burning, but the process is expensive and only partially successful, especially with organic sulfur. Alternatively, sulfur gases can be trapped by special devices ("scrubbers") in exhaust stacks, but again, the process is expensive (in terms of both money and energy) and not completely efficient. The burning of coal also produces a great volume of waste ash, amounting to 5 to 20 percent of the original volume of the coal, and this presents a further waste-disposal problem.

Coal mining is associated with additional concerns. Underground mines pose more hazards to the miners; after mining is complete, there is potential for subsidence of the surface as supporting timbers rot (recall figure 1.11), or even underground fires. Primarily for safety reasons, most coal mining in the United States is strip mining, and here again, the sulfur associated with the coal is problematic. The spoil banks of an unreclaimed strip mine are subject to rapid weathering, producing sulfuric acid runoff that is a water-pollution hazard and stunts regrowth of vegetation (figure 20.20A). Mine reclamation—involving regrading of spoils, restoration of a layer of topsoil, and replanting—can be effective (figure 20.20B) but costly, and with strip mines in dry areas, it can be difficult to reestablish vegetation.

Nuclear Power

Nuclear power actually comprises two different types of processes—*fission* and *fusion*—each with different advantages and limitations. Currently, only nuclear fission is commercially feasible.

Fission

Fission is the splitting apart of atomic nuclei into smaller ones, with the release of energy (figure 20.21). Very few isotopes (some 20 out of the more than 250 naturally occurring isotopes) can undergo fission spontaneously and do

A

B

Figure 20.20 Coal strip mining and mine reclamation. (*A*) Abandoned coal strip mine, Fulton County, Illinois. There is no lack of moisture, but acid runoff stunts plant growth. (*B*) Reclaimed portion of Indian Head coal strip mine, North Dakota, one year after reseeding.

(*A*) Photograph by Arthur Greenberg, courtesy EPA/National Archives. (*B*) Photograph by H. E. Malde, USGS Photo Library, Denver, Colorado

so in nature. Some additional nuclei can be induced to split apart, and the naturally fissionable nuclei can be made to split more rapidly, thus increasing the rate of energy release. The fissionable nucleus of most interest for modern nuclear power reactors is the isotope uranium-235.

A uranium-235 nucleus can be induced to undergo fission by the firing of another neutron into the nucleus. The nucleus splits into two lighter nuclei (not always the same two) and releases additional neutrons as well as energy. Some of the newly released neutrons can induce fission in other nearby uranium-235 nuclei, which, in turn, release more neutrons and more energy in breaking up, and so the process continues in a **chain reaction.** A controlled chain reaction, with continuous, moderate release of energy, is the basis for fission-powered reactors. The energy released heats cooling water that circulates through the reactor's core. The heat removed from the core is transferred, through a heat exchanger, to a second water loop in which steam is produced. The steam, in turn, is used to run turbines to produce electricity.

Only 0.7 percent of natural uranium is uranium-235. Natural uranium must be processed (enriched) to increase the concentration of this isotope to several percent of the total ura-

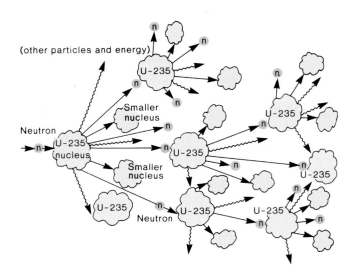

Figure 20.21 (Schematic) Nuclear fission and chain reaction involving uranium-235. Neutron capture by uranium-235 causes fission into two smaller nuclei plus additional neutrons, other subatomic particles, and energy. Released neutrons, in turn, cause fission in other uranium-235 nuclei.

nium to produce reactor-grade uranium capable of sustaining a chain reaction. As the reactor operates, the uranium-235 atoms are split and destroyed so that, in time, the fuel is so depleted ("spent") in this isotope that it must be replaced with a fresh supply of enriched uranium.

With the type of nuclear reactor currently in commercial operation in the United States, nuclear electricity-generating capacity in the United States probably could not be increased to much more than four times present

levels (to supply about 20 percent of total energy needs) by the year 2010 without serious shortages of uranium-235. However, uranium-235 is not the only possible fission-reactor fuel. When an atom of the nonfissionable uranium-238 absorbs a neutron, it is converted to plutonium-239, which *is* fissionable. During the chain reaction inside the reactor, as freed neutrons move about, some are captured by the abundant uranium-238 atoms, making plutonium.

Figure 20.26　Glen Canyon Dam hydroelectric project.

construction, including silting-up of reservoirs, habitat destruction and loss of farmland flooded by reservoirs, water loss by evaporation, and even earthquakes. Many potential sites—in Alaska, for instance—are too remote from population centers to be practical unless power transmission efficiency is improved. An alternative to the development of many new hydropower sites would be to add hydroelectric generating facilities to dams already in place for flood control, recreational use, or other purposes. Although the release of impounded water for power generation alters streamflow patterns, it is likely to be less disruptive than either the original dam construction or the creation of new dam/reservoir systems.

Like geothermal power, conventional hydropower is also limited by the stationary nature of the resource. In addition, hydropower is somewhat more susceptible to natural disruptions than other sources considered so far. Just as 100-year floods are rare, so are prolonged droughts, but they do happen.

For various reasons, then, it is unlikely that many new hydropower plants will be developed. This clean, economical, renewable energy source can continue indefinitely to make a modest contribution to energy use, but it cannot be expected to supply much more energy in the future than it now does.

Tidal Power

All large bodies of standing water on the earth, including the oceans and large lakes like the Great Lakes, have tides. Unfortunately, the energy represented by tides is too dispersed, in most places, to be useful. The difference in mean high-tide and low-tide water levels on the average beach is about 1 meter. A commercial tidal-power electricity-generating plant requires at least 8 meters difference between high and low tides for efficient generation of electricity, and a bay or inlet with a narrow opening that could be dammed to regulate water flow in and out. The proper conditions exist in very few places in the world. Tidal

power is being used at a small plant at Passamaquoddy, Maine, and also at locations in France, the Netherlands, and the former Soviet Union. Worldwide, the total potential of tidal power is estimated at only 2 percent of the energy potential of conventional hydropower, which is itself limited. Even where feasible, use of tidal power involves associated drawbacks, such as the cost and practical difficulty of constructing the necessary water-regulating system, and impacts on the bay ecology and perhaps on area fishing.

Wind Energy

Because the winds are ultimately powered by the sun, wind energy can be regarded as a derivative of solar energy. It is clean and, like sunshine, renewable indefinitely. Windmills currently are used for pumping groundwater and for the generation of electricity, usually on individual homesites.

Wind energy shares certain limitations with solar energy. It is dispersed, not only in two dimensions, but in three. Wind is also erratic and highly

Figure 20.27   Modern windmill array for power generation, Tehachapi Pass, California.
© Wm. C. Brown Communications, Inc./Photograph by Doug Sherman

variable in speed, both regionally and locally. The regional variations in potential power supply are even more significant than they might appear from comparing average wind velocities because windmill power generation increases as the cube of wind speed. So, if wind velocity doubles, power output increases by a factor of 8 (2 × 2 × 2). But most of the consistently windy places in the United States are rather far removed physically from most of the high-population and heavily industrialized areas. Even where average wind velocities are great, strong winds do not blow constantly. This presents a storage problem that has yet to be solved satisfactorily. In the near future, wind-generated electricity might most effectively be used to supplement conventionally generated power when wind conditions are favorable.

Most schemes for commercial wind-power generation of electricity involve *wind farms,* concentrations of many windmills in a few especially favorable sites (figure 20.27). The Great Plains region is the most promising area for initial sizeable efforts. The limits to wind-power use would then include the area that could be committed to windmill arrays, as well as the distance the electricity could be transmitted without excessive loss in the power grid. About 1,000 1-megawatt windmills would be required to generate as much power as a moderately large conventional electric power plant. The windmills have to be spread out so as not to block each other's wind flow. However, the land would not have to be devoted exclusively to windmills. Farming and the grazing of livestock, common activities in the Great Plains area, could con-

tinue on the land concurrently with power generation. The storage problem, however, would remain.

Biomass

The term **biomass** technically refers to the total mass of all the organisms living on earth. In an energy context, the term has become a catchall for various ways of deriving energy from organisms or from their remains.

The possibilities for biomass-derived fuels are many. Wood is a biomass fuel. There were eight wood-fueled electricity-generating plants in the United States in 1983, with a collective capacity ten times that of the wind-powered plants; by 1987, the plants fueled by burning wood and waste had a generating capacity nearly twenty times that of the wind-powered plants. Over 25 percent of U.S. house-holds burn some wood for heat, and 7.5 percent rely on wood as their primary heat source.

The production of alcohol from plant materials for use as fuel—either by itself or as an additive to gasoline—is another present use of biomass fuel. Sometimes, using biomass energy sources means burning waste plant materials after a crop is harvested. Because certain plants produce flammable, hydrocarbon-rich fluids, the possibility of "raising" liquid fuel on farms is being evaluated. A variation on this theme involves micro-organisms that produce oil-like substances: The organisms could be cultivated on ponds and their oil periodically skimmed off. Such possibilities represent potential future options.

A biomass fuel being utilized increasingly could be called "gas from garbage." Organic wastes, when decomposed in the absence of oxygen, yield a variety of gaseous products, among them methane (CH_4). The decay of organic wastes in sanitary landfill operations makes these landfills suitable sites for the production of methane. Gas straight from the landfill is too full of impurities to use alone, but landfill-derived gas can be blended with purer, pipelined natural gas to extend the gas supply. The city of Los Angeles is among several now making some use of landfill gas. Methane can also be produced by decay of manures, and on some livestock feedlots, this is proving to be a partial solution to energy-supply and waste-disposal problems simultaneously.

All biomass fuels are burned to release their energy. They are carbon-rich and thus share the carbon-dioxide-pollution problem of fossil fuels. However, unlike the fossil fuels, they have the very attractive feature of being renewable on a human time scale.

Summary

Economically valuable mineral deposits occur in a variety of geologic settings—igneous, metamorphic, and sedimentary. Valuable minerals may be found disseminated in igneous rocks, concentrated in plutons by gravity, crystallized as coarse-grained pegmatites, or deposited in hydrothermal veins associated with igneous activity. Sedimentary deposits include evaporites, placers, laterites, and other ores concentrated by weathering, and sediments and sedimentary rocks themselves (for example, limestone, sand and gravel, or sedimentary iron ores). Metamorphism plays a role in the formation of deposits of certain minerals, like graphite.

Both the occurrence of and demand for mineral and rock resources are very unevenly distributed worldwide. Projections for mineral use, even with conservative estimates of consumption levels, suggest that present reserves of most metals and other minerals could be exhausted within decades, both within the United States and globally. Price increases would lead to the development of some deposits that are presently uneconomic, but these are also limited. Strategies for averting further mineral shortages include applying new exploration methods to find more ores, looking to undersea mineral deposits not exploited in the past, and recycling metals to reduce the demand for newly mined material.

The world's major energy sources are the fossil fuels, formed from the remains of ancient plants and animals, modified by heat and pressure through time in the earth. All of the fossil fuels are nonrenewable energy sources. Petroleum and conventional natural gas deposits may be exhausted within decades. Coal resources could potentially last for centuries, but significant negative environmental impacts are associated with extensive coal use, including acid rain, land disturbance by mining activities, and the pollution and disposal problems posed by coal ash. Oil-shale deposits, containing the waxy hydrocarbon kerogen, are plentiful in the United States. However, the kerogen is thinly dispersed through the rocks, which would have to be strip-mined. Oil shale also requires a great deal of water to process, and most U.S. deposits occur in dry areas. U.S. tar-sand resources are negligible.

Conventional nuclear reactors consume the rare isotope uranium-235, the supply of which is severely limited. Significant expansion of the use of nuclear fission power would require the use of breeder reactors. All fission reactors present waste-disposal problems and raise concerns about radiation safety. Fusion power, which would be far cleaner, is technologically unfeasible for the present.

The inexhaustible or renewable energy sources include solar energy, geothermal power, hydropower, tidal and wind energy, and various biomass fuels. The principal practical applications of solar energy are for space heating and the generation of electricity; there is a storage problem in both contexts, and it is not yet possible to generate solar electricity efficiently for energy-intensive applications. Present use of geothermal energy, whether for hot-water heat or steam-generated electricity, is limited to a few sites along modern plate boundaries with recent volcanism. The thermal conductivity of rocks limits the lifetime of each individual geothermal field, typically to several decades. The potential for additional development of either conventional hydropower or tidal power is restricted by the number of suitable sites. Optimum sites for wind-energy development are mostly far removed from major population centers, and wind shares with sunlight the problem of variable power generation with changing weather. Biomass fuels can be replenished on a human time scale but share with fossil fuels the problem of carbon-dioxide pollution. There are significant technological, supply, or environmental limitations associated with each of the various alternative energy sources.

Terms to Remember

biomass	nonrenewable
breeder reactor	oil shale
chain reaction	ore
concentration	photovoltaic cells
(enrichment)	placer
factor	remote sensing
evaporite	renewable
Exclusive	reserves
Economic Zone	reservoir rocks
fission	resources
fossil fuels	spoil banks
fusion	strip mining
gasification	tailings
geothermal	tar sand
hydrothermal	traps
liquefaction	

Questions for Review

1. What is the distinction between *reserves* and *resources*? Under what conditions might some resources be reclassified as reserves?
2. What is an *ore*? How is its concentration factor defined?
3. Describe two kinds of magmatic ore deposits, and name one mineral mined from each.
4. What are *hydrothermal* ore deposits? Why are they especially associated with plate boundaries?
5. Under what conditions do evaporites form?
6. What is a *placer* deposit, and what kinds of minerals are concentrated in it?
7. Of which of the following does the United States use the largest quantity: iron, halite, platinum, or sand and gravel? Name one important application of each of these materials.
8. Name and describe two kinds of exploration techniques used in the search for mineral deposits.
9. How has the development of plate-tectonic theory aided in finding more ore deposits?
10. Is seawater a potential source of essential metals? Explain.
11. Describe one marine mineral resource, not presently exploited significantly, that might become important in the future.
12. Describe one potential environmental problem or hazard associated with (a) underground mining, (b) surface mining, and (c) mineral processing after mining.
13. What is an Exclusive Economic Zone? List three kinds of mineral deposits that might be encompassed by an EEZ.
14. What are the two basic initial requirements for forming a fossil-fuel deposit?
15. Briefly outline the process of petroleum formation and maturation.
16. Sketch or describe any two types of petroleum traps.
17. Approximately what proportion of estimated U.S. oil resources has been produced and consumed?
18. Describe any enhanced-recovery method for oil. What are the attractions of such methods?
19. From what materials is coal formed? How does coal's quality change with progressive heating?
20. Why is there interest in coal gasification and liquefaction processes?
21. Name and describe one environmental problem associated with (a) coal mining and (b) coal use/burning.
22. What is *oil shale*? How is fuel produced from it?
23. Cite three environmental or developmental problems associated with both oil shale and tar sand.
24. Compare and contrast nuclear fission and fusion processes.
25. If fission power is to be used extensively in the future, breeder reactors will be needed. Explain.
26. Cite two potential advantages of fusion power over fission. Why are fusion reactors not now in use?
27. For what two kinds of applications can solar energy make the most immediate contribution? What two principal limitations restrict the usefulness of solar energy for applications requiring a great deal of power, such as factories?
28. Briefly describe the nature of a natural geothermal system. Why are geothermal areas restricted geographically?
29. What are the principal applications of geothermal energy? What are two important limitations on its usefulness?
30. For what is conventional hydro-power now used? Discuss the extent to which additional hydro-power-generation sites might be developed.
31. What areas have the greatest potential for wind-power development? Name at least two factors that now limit the potential contribution of wind power.
32. Describe any two biomass fuel sources. What drawback do they share with the fossil fuels? What advantage do they have?

For Further Thought

1. Select one of the metals—such as aluminum, manganese, or tin—of which the United States controls very little or none of the identified world reserves. Investigate its uses in more detail; consider the implications of a cessation of imports. Has the United States any subeconomic deposits of this metal? Of what kind(s) are they, and how much metal do they contain relative to domestic demand?
2. Increased concern about energy conservation has led many homeowners to insulate their homes more and more snugly, to minimize heat loss by the escape of warmed air. Investigate the problems of "indoor air pollution" that have developed as a result.

3. Compare the cost estimates for shale oil, petroleum from tar sand, and gas from coal gasification with contemporary prices for conventional oil and natural gas. Do this for several different times—for example, 1972, 1977, 1982, and 1992. Determine the extent of oil and gas exploration and of development of the alternative fuels at the same times, and see what patterns, if any, emerge.

4. What energy sources are used by your local electric utility? Has the company changed its fuels, power-plant types, or plans for additional power plants over the last two or three decades? If so, to what extent have (a) economic factors and (b) environmental considerations entered into the decisions?

Suggestions for Further Reading

Agnew, A. F., ed. 1983. *International minerals: A national perspective.* Boulder, Colo.: Westview Press.

Brookins, D. G. 1981. *Earth resources, energy, and the environment.* Columbus, Ohio: Charles E. Merrill.

Deep ocean mining. 1982. *Oceanus* 25 (3).

Dick, R. A., and S. P. Wimpfen. 1980. Oil mining. *Scientific American* 243 (April): 182–88.

Fischett, M. A. 1986. The puzzle of Chernobyl. *IEEE Spectrum 23* (July): 34–41.

Guilmot, J. F., et al. 1987. *Energy 2000.* New York: Cambridge University Press.

Harris, D. P. 1984. *Mineral resource appraisal.* New York: Oxford University Press.

International Atomic Energy Agency, Vienna, Austria. (This agency publishes a quarterly bulletin on many aspects of nuclear power and other applications of radioactive materials.)

International Energy Agency. 1987. *Renewable sources of energy.* Paris: OCED.

Jensen, M. L., and A. M. Bateman. 1981. *Economic mineral deposits* 3d ed. New York: John Wiley and Sons.

Kendall, H. W., and S. J. Nadis, eds. 1980. *Energy strategies: Toward a solar future.* Cambridge, Mass.: Ballinger.

Macauley, G., L. R. Snowdon, and F. D. Ball. 1985. *Geochemistry and geological factors governing exploitation of selected Canadian oil-shale deposits.* Geological Society of Canada, Paper 85–13.

Money, L. J. 1984. *Transportation energy and the future.* Englewood Cliffs, N.J.: Prentice-Hall.

National Academy of Sciences. 1979. *Energy in transition, 1985–2010.* San Francisco: W. H. Freeman.

Petrick, A. 1986. *Energy resource assessment.* Boulder, Colo.: Westview Press.

Procter, R. M., G. C. Taylor, and J. A. Wade. 1984. *Oil and natural gas resources of Canada 1983.* Geological Society of Canada. Paper 83–31.

Ruedisili, L. C., and M. W. Firebaugh, eds. 1982. *Perspectives on energy* 3d ed. New York: Oxford University Press.

Sawkins, F. J. 1984. *Mineral deposits in relation to plate tectonics.* New York: Springer-Verlag.

Shusterich, K. M. 1982. *Resource management and the oceans.* Boulder, Colo.: Westview Press.

U.S. Bureau of Mines. 1991. *Mineral commodity summaries 1990.* (Published annually.)

U.S. Department of Energy, Energy Information Administration. (Publishes a wide variety of energy-related data. Sources used in this chapter include: *Annual energy outlook 1983* (published in 1984), *Annual energy review 1987* (1988), *Coal data book* (1980), *Commerical nuclear power: Prospects for the U.S. and the world* (1983), *Inventory of power plants in the U.S. 1983* (1984), *Nuclear plant cancellations: Causes, costs, and consequences* (1983), *Domestic uranium mining and milling industry* (1986), and *U.S. crude oil, natural gas, and natural gas liquids reserves* (1983).

Watson, J. 1983. *Geology and man.* Boston: Allen and Unwin.

Whitmore, F. C., Jr., and M. E. Williams, eds. 1982. *Resources for the twenty-first century.* U.S. Geological Survey Professional Paper 1193.

Zumberge, J. H. 1979. Mineral resources and geopolitics in Antarctica. *American Scientist* 67 (January): 68–76.

Mineral and Rock Identification

Mineral Identification

The following identification table lists many of the more common minerals. Representative chemical formulas are provided for reference. Some appear complex because of opportunities for solid solution; some have been simplified by limiting the range of compositions represented, although additional elemental substitutions are possible.

Following are a few general identification guidelines and comments:

1. Minerals showing metallic luster are usually sulfides (or native metals, but these are much rarer). Native metals have been omitted from the table; those few that are likely to be encountered, such as native copper or silver, may be identified by their resemblance to household examples of the same metals.
2. Of the nonmetals, the silicates are generally systematically harder than the nonsilicates. Hardnesses of silicates are typically over 5, with exceptions principally among the sheet silicates; many of the nonsilicates, such as sulfates and carbonates, are much softer.
3. Distinctive luster, cleavage, or other identifying properties are listed under the column "Other Characteristics." In a few cases, this column notes restrictions on the occurrence of certain minerals as a possible clue to identification; for example, "found only in metamorphic rocks" or "often found in pegmatites."

Rock Identification

One approach to rock identification is to decide whether the sample is igneous, sedimentary, or metamorphic and then look at the detailed descriptions in the corresponding chapters. How to identify the basic rock type? Here are some general guidelines:

1. Glassy or vesicular rocks are volcanic.
2. Coarse-grained rocks with tightly interlocking crystals are likely to be plutonic, especially if they lack foliation.
3. Coarse-grained sedimentary rocks differ from plutonic rocks in that the grains in the sedimentary rocks tend to be more rounded and interlock less closely. A breccia does have angular fragments, but the fragments in a breccia are typically rock fragments, not individual mineral crystals.
4. Rocks that are not very cohesive, that crumble easily into individual grains, are generally clastic sedimentary rocks. One exception would be a poorly consolidated volcanic ash, but this should be recognizable by the nature of the grains, many of which are glassy shards. (Note, however, that extensive weathering can make even a granite crumble.)
5. More cohesive, fine-grained sedimentary rocks may be distinguished from fine-grained volcanics because sedimentary rocks are generally softer and more likely to show a tendency to break along bedding planes. Phenocrysts, of course, indicate a (porphyritic) volcanic rock.
6. Foliated metamorphic rocks are distinguished by their foliation (schistosity, compositional banding). Also, rocks containing abundant mica, garnet, or amphibole are commonly metamorphic rocks.
7. Nonfoliated metamorphic rocks, like quartzite and marble, resemble their sedimentary parents but are harder, denser, and more compact; they may also have a glittery appearance on broken surfaces, due to recrystallization during metamorphism.

The foregoing guidelines are not infallible, but they should lead to the correct preliminary classification of most rocks commonly encountered.

Mineral	Formula	Color	Hardness	Other characteristics
amphibole	$(Na,Ca)_2(Mg,Fe,Al)_5Si_8O_{22}(OH)_2$	green, blue, brown, black	5 to 6	often forms needlelike crystals; two good cleavages forming 120-degree angle
apatite	$Ca_5(PO_4)_3(F,Cl,OH)$	usually yellowish	5	crystals hexagonal in cross section
azurite	$Cu_3(CO_3)_2(OH)_2$	vivid blue	3½ to 4	often associated with malachite
barite	$BaSO_4$	colorless	3 to 3½	high specific gravity, 4.5 (denser than most silicates)
beryl	$Be_3Al_2Si_6O_{18}$	aqua to green	7½ to 8	usually found in pegmatites
biotite	$K(Mg,Fe)_3AlSi_3O_{10}(OH)_2$	black	5½	excellent cleavage into thin sheets
bornite	Cu_5FeS_4	iridescent blue, purple	3	metallic luster
calcite	$CaCO_3$	variable; colorless if pure	3	effervesces in weak acid
chalcopyrite	$CuFeS_2$	brassy yellow	3½ to 4	
chlorite	$(Mg,Fe)_3(Si,Al)_4O_{10}(OH)_2$	light green	2 to 2½	cleaves into small flakes
cinnabar	HgS	red	2½	earthy luster; may show silvery flecks
covellite	CuS	blue	1½ to 2	metallic luster
dolomite	$CaMg(CO_3)_2$	white or pink	3½ to 4	powdered mineral effervesces in acid
epidote	$Ca_2FeAl_2Si_3O_{12}(OH)$	green	6 to 7	
fluorite	CaF_2	variable; often green or purple	4	cleaves into octahedral fragments; may fluoresce in ultraviolet light
galena	PbS	silver-gray	2½	metallic luster; cleaves into cubes
garnet	$(Ca,Mg,Fe)_3(Fe,Al)_2Si_3O_{12}$	variable; often dark red	7	glassy luster
graphite	C	dark gray	1 to 2	streaks like pencil lead
gypsum	$CaSO_4 \cdot 2H_2O$	colorless	2	
halite	$NaCl$	colorless	2½	salty taste; cleaves into cubes
hematite	Fe_2O_3	red or dark gray	5½ to 6½	red-brown streak regardless of color
kaolinite	$Al_2Si_2O_5(OH)_4$	white	2	earthy luster
kyanite	Al_2SiO_5	blue	5 to 7	found in high-pressure metamorphic rock; often forms bladelike crystals
magnetite	Fe_3O_4	black	6	strongly magnetic
malachite	$Cu_2CO_3(OH)_2$	green	3½ to 4	often forms in concentric rings of light and dark green
molybdenite	MoS_2	dark gray	1 to 1½	cleaves into flakes; more metallic luster than graphite
muscovite	$KAl_3Si_3O_{10}(OH)_2$	colorless	2 to 2½	excellent cleavage into thin sheets
olivine	$(Fe,Mg)_2SiO_4$	yellow-green	6½ to 7	glassy luster
phlogopite	$KMg_3AlSi_3O_{10}(OH)_2$	brown	2½ to 3	mica closely resembling biotite
plagioclase	$(Na,Ca)(Al,Si)_2Si_2O_8$	white to gray	6	may show fine striations on cleavage surfaces
potassium feldspar	$KAlSi_3O_8$	white; often stained pink	6	
pyrite	FeS_2	yellow	6 to 6½	metallic luster; black streak
pyroxene	$(Na,Ca,Mg,Fe,Al)_2Si_2O_6$	usually green or black	5 to 7	two good cleavages forming a 90-degree angle
quartz	SiO_2	variable; commonly colorless or white	7	glassy luster; conchoidal fracture
serpentine	$Mg_3Si_2O_5(OH)_4$	green to yellow	3 to 5	waxy or silky luster; may be fibrous
sillimanite	Al_2SiO_5	white	6 to 7	occurs only in metamorphic rocks; often forms needlelike crystals
sphalerite	ZnS	yellow-brown	3½ to 4	glassy luster
staurolite	$Fe_2Al_9Si_4O_{20}(OH)_2$	brown	7 to 7½	found in metamorphic rocks; elongated crystals may have crosslike form
sulfur	S	yellow	1½ to 2½	
sylvite	KCl	colorless	2	cleaves into cubes; salty taste, but more bitter than halite
talc	$Mg_3Si_4O_{10}(OH)_2$	white to green	1	greasy to slippery to the touch
tourmaline	$(Na,Ca)(Li,Mg,Al)(Al,Fe,Mn)_6(BO_3)_3Si_6O_{18}(OH)_4$	black, red, green	7 to 7½	elongated crystals, triangular in cross section; conchoidal fracture

A key to aid in rock identification

Igneous rocks

I. Extremely coarse-grained: Rock is a pegmatite. (Most pegmatites are granitic, with or without exotic minerals.)

II. Phaneritic (coarse enough that all grains are visible to the naked eye).
   - A. Significant quartz visible; only minor mafic minerals: *granite*.
   - B. No obvious quartz; feldspar (light-colored) and mafic minerals (dark) in similar amounts: *diorite*.
   - C. No quartz; rock consists mostly of mafic minerals: *gabbro*.
   - D. No visible quartz *or* feldspar: Rock is ultramafic.

III. Porphyritic with fine-grained groundmass: Go to part IV to describe groundmass (using phenocryst compositions to assist); adjective "porphyritic" will preface rock name.

IV. Aphanitic (grains too fine to distinguish easily with the naked eye).
   - A. Quartz is visible or rock is light in color (white, cream, pink): probably *rhyolite*.
   - B. No visible quartz; medium tone (commonly gray or green); if phenocrysts are present, commonly plagioclase, pyroxene, or amphibole: probably *andesite*.
   - C. Rock is dark, commonly black; any phenocrysts are olivine or pyroxene: *basalt*.
   - D. Rock is glassy and massive: *obsidian* (regardless of composition).
   - E. Rock consists of gritty mineral grains, ash and glass shards: *ignimbrite* (welded tuff).

Sedimentary rocks

I. Rock consists of visible shell fragments or of oolites: *limestone*.

II. Rock consists of interlocking grains with texture somewhat like that of igneous rock and is light in color: probable chemical sedimentary rock.
   - A. Tastes like table salt: *halite*.
   - B. No marked taste; hardness of 2 (if grains are large enough to scratch); does not effervesce: *gypsum*.
   - C. Effervesces in weak HCl: *limestone* (calcite).
   - D. Effervesces weakly in HCl, only if scratched: *dolomite*.

III. Rock consists of grains apparently cemented or compacted together: probable clastic sedimentary rock.
   - A. Coarse grains (several millimeters or more in diameter), perhaps with a finer matrix: *conglomerate* if the grains are rounded, *breccia* if they are angular.
   - B. Sand-sized grains; gritty feel: *sandstone*. If predominantly quartz grains, *quartz sandstone;* if roughly equal proportions of quartz and feldspar, *arkose;* if many rock fragments, and perhaps a fine-grained matrix, *greywacke*.
   - C. Grains too fine to see readily with the naked eye: *mudstone*. If rock shows lamination, and a tendency to part along parallel planes, *shale*.

IV. Relatively dense; compact, dark, no visible grains; massive texture, conchoidal fracture: *chert* (silica).

Metamorphic rocks

I. Nonfoliated; compact texture with interlocking grains: identified by predominant mineral(s).
   - A. If quartz-rich, perhaps with a sugary appearance: *quartzite*.
   - B. If calcite or dolomite (identified by effervescence, hardness): *marble*.
   - C. Rock consists predominantly of amphiboles: *amphibolite*.

II. Foliated: classified mainly by texture.
   - A. Very fine-grained; pronounced rock cleavage along parallel planes, to resemble flagstones: *slate*.
   - B. Fine-grained; slatelike, but with glossy cleavage surfaces: *phyllite*.
   - C. Coarser grains; obvious foliation, commonly defined by prominent mica flakes, sometimes by elongated crystals like amphiboles: *schist*.
   - D. Compositional or textural banding, especially with alternating light (quartz, feldspar) and dark (ferromagnesian) bands: *gneiss*.

Glossary

Aa   Rough, blocky lava flow.

ablation   Loss of material from a glacier, by melting, evaporation, or calving.

abrasion   Grinding erosion by rocks entrained in glacial ice or by windblown sand.

absolute age   Old name for *radiometric age* (or date).

abyssal hills   Low hills, several hundred meters high, found in the abyssal plains.

abyssal plains   The flat areas of the deep ocean basins.

accreted terrane   A suspect terrane that is found to have been transported some distance prior to accretion onto a continental margin.

acid rain   Rainfall that is more acidic than typical precipitation, especially as due to sulfur pollution in the atmosphere (forming sulfuric acid).

active margin   A continental margin at which there is significant volcanic and earthquake activity; commonly, a convergent plate margin.

active volcano   Volcano that is fresh-looking and that has erupted within recent history.

aftershocks   Smaller earthquakes that follow a major earthquake.

A horizon   The topmost soil horizon, including topsoil; also known as the *zone of leaching.*

alluvial fan   A wedge-shaped sediment deposit left where a tributary flows into a more slowly flowing stream or where a mountain stream flows into a desert.

alluvium   Stream-deposited sediment.

alpine glacier   A small glacier found in a mountainous region; commonly, a valley glacier.

amphiboles   Hydrous, ferromagnesian, double-chain silicates.

amphibolite   Metamorphic rock rich in amphiboles.

andesite   Volcanic rock intermediate in composition.

angle of repose   The steepest angle at which a slope of unconsolidated material is stable.

angular unconformity   Unconformity at which bedding of rocks above and below is oriented differently.

anion   A negatively charged ion.

anticline   An antiform in which the oldest beds are at the center of the fold.

antiform   An arching fold, in which limbs dip away from the axis.

aphanitic   Having crystals too small to be seen with the naked eye.

aquiclude   Rock that is effectively impermeable on a human time scale.

aquifer   Rock sufficiently porous and permeable to be useful as a source of water.

aquitard   Rock of low permeability, through which water flows very slowly.

arête   Sharp-spined ridge left by erosion of parallel valley glaciers to either side of the ridges.

artesian   System in which groundwater in a confined aquifer is under extra hydrostatic pressure, so that the water can rise above the aquifer containing it.

ash   Fine pyroclastic material, with fragments up to about 2 millimeters in diameter.

assimilation   Process by which magma incorporates and melts bits of country rock.

asthenosphere   Weak, plastic, partly molten layer of the upper mantle directly below the lithosphere.

atom   The smallest particle into which a chemical element can be subdivided.

atomic mass number   The sum of the number of protons and number of neutrons in a particular atomic nucleus.

atomic number   The number of protons characteristic of a particular element.

aureole   Contact-metamorphic zone around a pluton.

axial surface   Surface—often planar—dividing the two limbs of a fold.

axial trace   The intersection of the axial plane of a fold with the land surface.

axis   Line of intersection of the axial plane with the surface of a fold.

backwash   Return flow of swash to the sea.

bankfull stage   The condition in which stream stage just equals stream-bank elevation.

barchan dune   A crescent-shaped transverse dune, with arms pointing downwind.

barrier island   Long, low, narrow island parallel to a coast.

basalt   Mafic volcanic rock; volcanic equivalent of gabbro.

base flow   Streamflow supported by groundwater in adjacent rock or soil.

base level   Ordinarily, the level of the water surface at a stream's mouth; the lowest level to which a stream can cut down.

basin   A synform in which rocks dip inward on all sides; also, any depression in which sediments are deposited.

batholith   A massive, discordant pluton, often produced by multiple intrusions.

beach   A gently sloping shore covered with sediment and washed by waves and tides.

beach face   That portion of the beach exposed to direct wave and swash action.

bedding   Layering, as in sedimentary rocks.

bed load   The amount of material moved by a stream along its bed, by rolling or by saltation.

bedrock geology   The geology as it would appear with overlying soil and vegetative cover stripped away.

Benioff zone   Zone of earthquake foci dipping into the mantle away from a trench, resulting from subduction of lithosphere.

berm   A flat or gently sloping zone behind a beach face.

B horizon   Soil layer at intermediate depth; also known as *zone of accumulation.*

biomass   Fuel derived from modern organisms.

block Pyroclastic fragment coarser than cinders.

blowout Bowl-shaped depression hollowed out of sand by deflation.

body waves Seismic waves that travel through the earth's interior (P waves and S waves).

bomb Pyroclastic fragment, formed from an erupted blob of magma, that may take on a streamlined shape during flight.

Bowen's reaction series Predicted sequence of crystallization of principal silicates from a magma.

braided stream A stream with multiple channels that divide and rejoin.

breccia Clastic sedimentary rock consisting of angular fragments in a finer-grained matrix.

breeder reactor Nuclear fission reactor which, in addition to producing energy, produces new fissionable material for future use as fuel.

brittle Describes material that tends to rupture rather than to deform under stress.

caldera A large bowl-shaped summit depression in a volcano; may be formed by explosion or collapse.

caliche Hard crust of calcium carbonate formed in soils in arid regions.

calving Formation of icebergs as chunks of ice break off a glacier that terminates in water.

capacity The maximum total load a given stream can move.

capillary action Movement of water toward drier soil through fine pores in the soil and along the grain surfaces.

carbonate Nonsilicate mineral containing carbon and oxygen in the proportions of one atom of carbon to three atoms of oxygen (CO_3).

cast Replica of a fossil form created by the filling of a *mold.*

catastrophism A now-discredited theory that explained earth's history as a static one, punctuated by global catastrophes that were the only agents of change.

cation A positively charged ion.

cementation Process by which sediments are stuck together through the deposition of mineral material between grains.

Cenozoic The most recent era of geologic time, from 66 million years ago to the present.

chain reaction A sequence of fission events in which each event triggers additional fission events.

chain silicates Silicates in which silica tetrahedra are linked in one dimension by the sharing of oxygen atoms.

channelization Modification of a stream channel—for example, by straightening of meanders or dredging of the channel to deepen it.

chemical maturity Measure of the extent to which sediment has been depleted in soluble or easily weathered minerals.

chemical sediment Sediment precipitated directly from solution.

chemical weathering Solution or chemical breakdown of minerals by reaction with water, air, or dissolved substances.

chilled margin Fine-grained rock at the margin of a pluton, showing the effects of rapid cooling.

C horizon The deepest soil layer, consisting mainly of coarse chunks of bedrock.

cinder cone A volcanic structure built of pyroclastic materials.

cinders Glassy, bubbly, pyroclastic material that falls to the ground as solid fragments.

cirque A bowl-shaped depression formed at the head of an alpine glacier.

clastic Describes rock or sediment made of fragments of preexisting rocks and minerals.

cleavage (mineral) Tendency of a mineral to break preferentially along planes in certain directions in the crystal structure.

cleavage (rock) Tendency of rock to break along parallel planes, corresponding to planes along which platy minerals are aligned.

closed system A system that neither gains nor loses matter; an isolated system with respect to mass transfer.

coal Solid fossil fuel formed from the remains of land plants.

coastline Zone at which land and water meet; also, the geometry of this zone.

columnar jointing Development of polygonal columns in a lava flow during cooling.

compaction Compression and consolidation of sediment under compressive stress.

competence The largest size of particle a stream can move as bed load.

composite volcano A volcanic cone formed of interlayered lava flows and pyroclastics.

compound A chemical combination of two or more elements, in specific proportions, with a distinct set of physical properties.

compressive stress Stress tending to compress or squeeze an object.

concentration (enrichment) factor Enrichment of an ore in a metal of interest, relative to that metal's concentration in average crustal rock.

concordant Having contacts parallel to the structure in adjacent rocks.

cone of depression Conical depression of the water table or potentiometric surface caused by pumped extraction of groundwater.

confined aquifer An aquifer overlain by an aquiclude or aquitard.

confining pressure Directionally uniform pressure to which rocks at depth are subjected.

conglomerate Clastic sedimentary rock consisting of rounded fragments in a finer-grained matrix.

contact metamorphism Metamorphism characteristic of wallrocks surrounding a pluton that are subjected to locally increased temperature only.

continental drift The concept that the continents have shifted in position over the earth; also, the process by which this has occurred.

continental glacier A thick, extensive ice cap or ice sheet covering a significant portion of a continent.

continental rise Gently sloping region between the foot of the continental slope and the abyssal plains.

continental shelf The nearly level, shallowly submerged zone immediately offshore from a continent; water depths on the shelf are typically less than 100 meters.

continental slope The continental-marginal zone extending from the outer edge of the continental shelf down to the more gently sloping ocean depths (continental rise or abyssal plains).

continuous reaction series Crystallization series in which early crystals react with the melt without changes in mineralogy; the plagioclase branch of Bowen's reaction series.

contour interval The difference in elevation represented by two successive contours on a topographic map.

contour line A line joining points of equal elevation on a map.

convection cell Circulating mass of material (in air, water, or asthenosphere) in which warm material rises, moves laterally, cools, sinks, and is reheated, cycling back to rise again.

convergent plate boundary Plate boundary at which two plates are moving together: subduction zone or zone of continental collision.

coral atoll A ring-shaped coral reef structure, formed around an island that is now submerged or eroded away.

core The innermost, iron-rich zone of the earth; the outer core is molten, the inner core solid.

correlation Determination that two or more distinct rock units are of the same age and/or related in origin.

cosmic abundance curve Graph depicting the relative abundances of the chemical elements as a function of atomic number.

country rock Rock into which a pluton is intruded; also known as wallrock.

covalent bond A bond formed by the sharing of electrons between atoms.

craton Stable continental interior.

creep (fault) Slow, gradual, more-or-less continuous slippage along a fault zone.

creep (mass movement) Very slow mass movement, not noticeable during direct observation.

crest (flood) The maximum stage reached during a flood event.

crest (wave) The highest point on a wave.

crevasses Deep, vertical cracks in brittle glacier ice.

cross-bedding Sequence of inclined sedimentary beds deposited by flowing wind or water.

cross section (geologic) Interpretation of geology and structure in the third (vertical) dimension based on rock exposures and attitudes at the surface; drawn in a particular vertical plane.

crust The outermost compositional shell of earth, 10 to 40 kilometers thick, consisting predominantly of relatively low-density silicates.

crystal form External shape of crystals; distinguished from internal crystal structure.

crystalline Describes a solid having a regular, repeating, symmetric arrangement of atoms.

Curie temperature Temperature above which a magnetic material loses its magnetization; different for each such material.

daughter nucleus Nucleus produced by radioactive decay.

debris avalanche A flow involving a wide range of types and sizes of material.

deep layer The deepest water in the oceans, with temperatures often near freezing; circulates very slowly.

deflation Removal of sediment by wind.

delta A sediment wedge deposited at a stream's mouth.

dendritic drainage Irregular, branching drainage pattern.

desert A region having so little vegetation that it is incapable of supporting a significant population.

desertification Rapid conversion of marginally habitable arid land into true desert; typically accelerated by human activities.

desert pavement A surface of coarse rocks protecting finer sediment below; formed by selective removal of fine surficial material.

dewatering Release of water from pores and/or breakdown of hydrous minerals under conditions of increasing pressure or temperature.

diagenesis Set of processes by which lithification is accomplished; occurs at lower temperatures than metamorphism.

dike A tabular, discordant pluton.

diorite Plutonic rock of intermediate composition, consisting of ferromagnesian minerals and feldspar, with little quartz and no olivine.

dip The angle made by a line or plane with the horizontal.

dip-slip fault Fault along which movement is parallel to dip.

directed stress Stress that is not uniformly intense in all directions.

discharge The volume of water flowing past a given point in a specified period of time; equal to flow velocity multiplied by the cross-sectional area of the channel.

disconformity An unconformity at which the bedding of rocks above and below is parallel.

discontinuous reaction series Crystallization sequence in which early crystals are transformed into new minerals by reaction with the melt, with abrupt changes in crystal structure; the ferromagnesian branch of Bowen's reaction series.

discordant Having contacts that cut across or are set at an angle to the structure of the adjacent rocks.

dissolved load The quantity of material carried in solution by a stream.

divergent plate boundary Plate boundary at which plates are moving apart: ocean spreading ridge or continental rift zone.

dolomite A carbonate mineral, $CaMg(CO_3)_2$, or the chemical sedimentary rock made predominantly of that mineral.

dolostone Chemical sedimentary rock composed predominantly of the mineral dolomite.

dome An antiform dipping radially in all directions.

dormant Describes a volcano presently inactive but believed capable of future eruption.

downcutting The downward erosion by a stream toward its base level.

drainage basin The area from which a stream system draws its water.

drift Sediment that has been transported and deposited by glacial action.

drowned valley A stream or glacial valley partially flooded by seawater; occurs on a submergent coastline.

drumlins Elongated mounds of till oriented parallel to ice flow.

dune A low mound or ridge of sediment deposited by wind.

dynamic equilibrium Condition in which two opposing processes are in balance; for a stream, the condition in which erosion and deposition in the channel are equal.

earthquake Ground displacement associated with the sudden release, in the form of seismic waves, of built-up stress in the lithosphere.

elastic deformation Deformation in which strain is proportional to applied stress and the material returns to its original dimensions when the stress is released.

elastic limit The stress beyond which material no longer behaves elastically.

elastic rebound The phenomenon whereby stressed rocks behave elastically before and after an earthquake, returning afterward to an undeformed, unstressed condition.

electron A negatively charged subatomic particle, found outside the atomic nucleus.

element The simplest kind of chemical substance; elements cannot be decomposed further by chemical or physical means.

end moraine A ridge of till accumulated at the end of a glacier.

enhanced recovery Any method for increasing the proportion of accumulated oil and gas extracted from a petroleum reservoir.

eolian Formed by or related to wind action.

ephemeral stream A stream that flows only occasionally in direct response to precipitation.

epicenter The point on the earth's surface directly above an earthquake's focus.

era Major subdivision of the geologic time scale.

erratic An isolated large boulder not derived from local bedrock; a depositional feature of glaciers.

esker A winding ridge of till deposited by a stream flowing in and under a melting glacier.

estuary A coastal body of brackish water, open to the sea.

eustatic Describes simultaneous worldwide rise or fall of sea level.

evaporite A mineral deposit formed by evaporation of seawater in a restricted basin; also, the minerals of such deposits.

Exclusive Economic Zone Territory extending up to 200 miles out from a nation's shoreline, within which that nation has exclusive rights to exploit marine resources.

exfoliation Breakup of exposed plutonic rocks in concentric sheets due to release of stress by unloading.

extinct Describes a volcano expected never to erupt again.

fall A free-falling mass movement in which the moving mass is not always in contact with the ground below.

fault A planar break in rock along which there is movement of one side relative to the other.

fault breccia Breccia formed by mechanical breakup of rocks during displacement along a fault zone.

feldspars One group of framework silicates, containing aluminum and calcium, sodium, or potassium; collectively, the most abundant minerals in the crust.

felsic Rock rich in feldspar and silica (quartz).

ferromagnesian Silicate containing significant iron and/or magnesium.

firn Dense, coarsely crystalline snow partially converted to ice.

fission Process by which an atomic nucleus is split into two or more smaller nuclei.

fissure eruption Eruption of lava from a crack rather than from a pipelike vent.

floodplain Nearly flat area around a stream channel, into which the stream overflows during floods.

flood stage The condition in which stream stage is above channel bank elevation, so that the stream overflows its banks.

flow A mass movement of unconsolidated material in which the material moves in a chaotic or disorganized fashion, rather than as a coherent unit.

fluid injection A proposed means of increasing pore fluid pressure and decreasing shear strength along locked faults to release built-up stress.

focus The point of first break along a fault during an earthquake.

foliation A texture, usually metamorphic, involving parallel alignment of linear or planar minerals, or compositional banding.

fossil Remains or evidence of ancient life.

fossil fuels Any of the carbon-rich fuels produced through heat, pressure, and time from the remains of organisms.

fractional crystallization Crystallization of magma with early crystals removed or isolated from later reaction with the remaining melt.

fracture Irregular breakage; contrasted with *cleavage*.

framework silicate A silicate in which silica tetrahedra are linked in three dimensions by shared oxygen atoms.

frost wedging Breakup of rock by the expansion of water freezing in cracks.

fumarole Steam and gas vent caused by subsurface water being heated by shallow magma or hot rock.

fusion Process by which two or more nuclei are combined to form a larger nucleus.

gabbro A mafic plutonic rock rich in ferromagnesians and plagioclase feldspar.

gasification Process by which coal is converted to a gaseous hydrocarbon fuel.

geosyncline A large syncline, of regional scale.

geothermal Related to the heat of the earth's interior; use of geothermal energy involves extraction of that heat through circulating subsurface water.

geothermal gradient The increase in temperature with depth in the earth.

geyser A feature characterized by intermittent ejection of hot water and steam, heated by shallow magma or hot rocks.

glacier A mass of ice, on land, that moves under its own weight.

glass A solid lacking a regular crystal structure, in which atoms are randomly arranged.

gneiss A metamorphic rock showing banded texture, usually defined by differences in mineralogy between bands.

graben A downdropped block bounded by steeply dipping faults.

graded bedding Vertical progression of grain sizes within a sediment layer, from coarse to fine, or vice versa.

graded stream A stream in dynamic equilibrium.

gradient Steepness or slope of a stream channel along its length.

granite A plutonic rock rich in quartz and potassium feldspar.

greenhouse effect Atmospheric heating resulting from the trapping of heat by carbon dioxide and other gases in the atmosphere.

groundmass The finer-grained matrix of a porphyritic rock.

ground moraine Sheet of moraine left by a melting glacier.

ground truth Check of the interpretation of remotely sensed data by direct contact (observation on the ground).

groundwater Water in the saturated zone, below the water table.

gullying Formation by water of large erosional channels on a sloping soil surface.

guyot A flat-topped seamount.

half-life The length of time required for half of an initial quantity of a radioisotope to decay.

halide A nonsilicate containing a halogen element (Cl, F, Br, I).

hanging valley The valley of a tributary glacier; smaller than the main glacier valley, with a higher floor.

hardness The ability to resist scratching; measured on the Mohs scale of relative hardness.

hard water Water containing high concentrations of dissolved calcium, magnesium, and iron.

headward erosion The cutting back of a stream channel at its source.

height (wave) The difference in elevation between a wave's crest and trough.

hinge The most sharply curved part of a fold.

horn A peak formed by headwall erosion by several alpine glaciers diverging from the same topographic high.

horst An uplifted block bounded by high-angle faults.

hot spot An isolated area of active volcanism not associated with a plate boundary.

hot springs Springs heated by shallow magma bodies or young, hot rocks.

hydrograph Graph of stream stage or discharge as a function of time.

hydrologic cycle The cycle of precipitation, evaporation, infiltration, and migration of the water in the hydrosphere.

hydrosphere All water that is at and near the earth's surface and not chemically bound in rocks.

hydrostatic pressure Fluid pressure.

hydrothermal Literally, "hot water"; describes processes or ore deposits related to circulating subsurface water warmed by shallow magma or hot rock; hydrothermal fluids commonly contain dissolved minerals and gases.

hydrothermal vents Areas along spreading-ridge systems where waters heated by reaction with new lithosphere emerge into the colder ocean.

hydrous Containing water or hydroxyl (OH⁻) ions.

hypothesis A conceptual model or explanation for a set of data, measurements, or observations.

ice age A period of very extensive continental glaciation; when capitalized ("Ice Age"), it refers to the last such episode, 2 million to 10,000 years ago.

ice wedging See *frost wedging*.

igneous Formed from or related to magma.

incised meanders Meanders cut deeply into rock, with little or no floodplain at channel level.

index fossils Fossils that are particularly useful for correlation.

index minerals Minerals stable over a restricted range of pressure and/or temperature conditions; useful in evaluating metamorphic grade attained.

inert Not tending to bond or form compounds with other elements.

infiltration Percolation of water into the ground.

intensity The size of an earthquake as measured by its effects on structures; one earthquake may have several intensities that decrease with increasing distance from the epicenter.

internal drainage Stream drainage into an enclosed, landlocked basin.

ion An electrically charged atom.

ionic bond A bond formed by attraction between cations and anions.

island arc A line or arc of volcanic islands formed over, and parallel to, a subduction zone overlain by oceanic lithosphere.

isograd A line on a map connecting points of equal metamorphic grade, as determined by index minerals.

isostasy The tendency of crust and lithosphere to float at an elevation consistent with the density and thickness of the crustal rocks relative to underlying mantle.

isostatic equilibrium Condition in which the mass of rock above a given level in the earth is everywhere the same.

isotopes Atoms of the same chemical element that differ in numbers of neutrons in the nucleus.

joint A planar break in rock without relative movement of rocks on either side of the break.

joint set A set of parallel joints.

karst topography Topography characterized by abundant sinkholes and other solution features.

kettle A hole in glacial outwash, formerly occupied by a block of stranded ice.

knickpoint An abrupt change in streambed elevation—for example, at a waterfall.

laccolith A concordant pluton with flat bottom and domed country rock above.

lahar A volcanic mudflow.

lamination Very fine or thin bedding.

landslide Any rapid mass movement; contrasted with *creep*.

lateral moraine Moraine deposited at the sides of a valley glacier.

lateritic soil Extensively leached soil; characteristic of tropical climates.

lava Magma that flows out at the earth's surface.

law A basic concept or mathematical relationship that is invariably found to be true.

Law of Faunal Succession The concept that life forms change through time, each specific form corresponding to a unique period of earth history.

leaching The removal of soluble chemicals by infiltrating or percolating water.

levees Ridges along the bank of a stream; may be natural or artificial.

limbs The two sides of a fold, on either side of its axial plane.

limestone A carbonate-rich (especially calcite-rich) chemical sedimentary rock.

liquefaction (coal) Process by which coal is converted to liquid hydrocarbon fuel.

liquefaction (earthquake) Quicksand-like condition with loss of soil strength; occurs when water-saturated soil is shaken by seismic waves.

lithification The conversion of sediment into rock.

lithosphere The rigid outermost layer of the earth, 50 to 100 kilometers thick, encompassing the crust and uppermost mantle.

littoral drift Sand movement along the length of a beach that occurs in the presence of longshore currents.

load The total amount of material moved by a stream.

locked fault An active fault along which friction is preventing stress release through creep; no displacement is occurring.

loess Silt-sized sediment deposited by wind.

longitudinal dunes Dunes elongated parallel to the direction of wind flow.

longitudinal profile A diagram of the elevation of a streambed along its length.

longshore current Net current parallel to a coastline, caused when waves approach the shore at an oblique angle.

lopolith A concordant pluton with a floor that is concave upward.

low-velocity layer Zone within the upper mantle characterized by lower seismic velocities than layers immediately above and below it, a result of plastic behavior and/or partial melting of rocks in this zone.

luster The surface sheen exhibited by a mineral.

mafic A rock, magma, or mineral rich in iron and magnesium.

magma A silicate melt, usually containing dissolved volatiles, sometimes also containing crystals.

magma mixing Process by which two compositionally dissimilar magmas are combined into one.

magnitude The size of an earthquake as measured by vertical ground displacement near the epicenter.

manganese nodules Lumps of manganese and iron oxides and hydroxides, along with other metals, found on the sea floor.

mantle The zone of the earth's interior between crust and core; rich in ferromagnesian silicates.

map unit A distinct, identifiable rock unit used in preparing a geologic map.

marble Metamorphosed limestone.

mass wasting Downslope movement of material under the influence of gravity.

matrix Fine-grained sediment filling spaces between coarse grains in poorly sorted clastic sediment or rock.

meanders Lateral bends in a stream channel.

mechanical weathering Physical breakup of rocks, without changes in their composition.

medial moraine Moraine formed by the joining of lateral moraines as tributary glaciers flow into a valley glacier.

Mesozoic The middle era of the Phanerozoic; the time from about 245 to 66 million years ago.

metamorphic grade A measure of the intensity of metamorphism to which a metamorphic rock was subjected.

metamorphism Literally, "change in form" of rocks, brought about particularly through the application of heat and pressure.

metasomatism The introduction of ions in solution into a rock, and the resulting alteration of that rock.

micas A group of sheet silicates characterized by the tendency to cleave well between sheets of silica tetrahedra.

migmatite "Mixed rock," partly melted during metamorphism, having a mix of igneous and metamorphic characteristics.

milling Erosion by water-borne sediment.

mineral A naturally occurring, inorganic, solid element or compound, with a definite composition or compositional range and a regular internal crystal structure.

mineralogical maturity The extent to which a sediment has been depleted in easily weathered minerals and enriched in resistant ones.

mineraloid A material satisfying the definition of a mineral except that it lacks a regular internal crystal structure.

Mohorovičić discontinuity (Moho) The boundary between crust and mantle.

mold An impression of a fossil form preserved in rock or sediment.

moraine A landform composed of till.

mouth Where a stream reaches its base level or terminates.

mud cracks Cracks in fine sediment caused by shrinkage during dehydration.

mudstone A very fine-grained clastic sedimentary rock such as siltstone or claystone.

native element A mineral consisting of a single chemical element.

neap tides Least extreme tides, which occur when sun and moon are at right angles relative to the earth.

neck (volcanic) A pipelike, discordant pluton.

neutron An electrically neutral subatomic particle, generally found in an atomic nucleus and having approximately the same mass as a proton.

nonrenewable Describes a resource that is not being produced at a rate comparable to that at which it is being consumed.

normal fault A dip-slip fault in which the hanging wall moves downward relative to the footwall.

nucleus Central unit of an atom, containing protons and neutrons.

nuée ardente A hot, glowing cloud of volcanic ash and gas, so dense that it flows down the volcano's slopes.

obduction Process by which a segment of lithosphere is placed atop another.

oblique-slip fault A fault involving both strike-slip and dip-slip movements.

obsidian Massive volcanic glass.

oil shale Sedimentary rock containing a waxy, solid hydrocarbon, kerogen.

olivine A ferromagnesian silicate with structure consisting of individual silica tetrahedra.

ooze Fine-grained, water-rich, siliceous or calcareous pelagic sediment of biogenic origin.

ophiolite Complex assemblage of marine sediments, mafic and ultramafic rocks, found on a continent; generally believed to be a piece of obducted oceanic lithosphere.

order (stream) Hierarchical rank of a stream based on the number of levels of tributaries: a first-order stream has no tributaries, a second-order stream has only first-order tributaries, and so on.

ore Rock in which a valuable or useful metal occurs in sufficient concentration to be economic to mine.

orogenesis Set of plutonic, metamorphic, and tectonic processes involved in mountain-building.

orogeny The period of time over which orogenesis occurs in a particular mountain range.

outwash Glacial sediment moved and redeposited by melting ice.

outwash plain Broad sheet of outwash deposited in front of a glacier by the collective action of many meltwater streams.

overbank deposits Alluvium deposited outside a stream channel during flooding.

oxbows Cut-off meanders.

oxide A nonsilicate containing oxygen with one or more metals.

pahoehoe Lava flow with a smooth, ropy appearance.

paleomagnetism "Fossil magnetism" preserved in rocks, reflecting orientation of the earth's magnetic field at the time magnetization was acquired.

Paleozoic The oldest era in the Phanerozoic; the time from approximately 570 to 245 million years ago.

Pangaea Single supercontinent that existed approximately 200 million years ago.

parabolic dunes Crescent-shaped dunes with arms pointing upwind.

parent nucleus (radioactive) An unstable, decaying nucleus.

parent rock Original rock from which a metamorphic rock was formed.

partial melting Melting of only a portion of a rock.

passive margin A geologically or tectonically quiet continental margin, lacking significant volcanic or seismic activity.

pedalfer A moderately leached soil rich in iron and aluminum oxides and hydroxides.

pediment A gently sloping bedrock surface at the foot of mountains bordering a desert.

pedocal A soil retaining many of its soluble minerals, especially calcite.

pegmatite Very coarse-grained igneous rock.

pelagic sediments Fine-grained sediments of the open ocean.

period (time) Subdivision of an era of the geologic time scale.

period (wave) The time interval between passage of two successive wave crests or troughs by a fixed point.

periodic table A regular arrangement of chemical elements that reflects patterns of chemical behavior related to the electronic structure of atoms.

permeability A measure of the ease with which fluids move through rocks or sediments.

phaneritic Having crystals large enough to be distinguished with the naked eye.

Phanerozoic Major division (eon) of the geologic time scale, 570 million years ago to the present, spanning the time over which complex life forms have been abundant.

phase change A change in mineralogy, crystal structure, or physical state, with no gain or loss of chemical elements.

phenocryst A coarse crystal in a porphyritic rock.

photovoltaic cell Device for converting solar energy to electricity.

phreatic eruption A volcanic steam explosion caused when magma heats subsurface water.

phreatic zone The zone of saturation, in which pores in rock or soil are filled with water.

phyllite A fine-grained metamorphic rock, formed by progressive metamorphism of slate, in which cleavage planes shine with light reflected from small mica flakes.

physical geology That branch of geology concerned particularly with the materials of the earth and the physical processes that shape it.

pipe See *neck*.

placer Sedimentary ore deposit formed when dense, weathering-resistant minerals are concentrated by the action of flowing water.

plastic Behavior in which deformation is not proportional to applied stress and is permanent; the material stays deformed when stress is removed.

plateau (oceanic) A broad topographic high, shallowly submerged or slightly exposed, within the ocean basins; often underlain by continental-type crust.

plate tectonics The theory according to which the lithosphere is broken up into a series of rigid plates that can move over the earth's surface.

platform That part of the continental craton consisting of nearly flat-lying, relatively undeformed sedimentary rocks.

playa A "dry lake," floored by fine sediment, formed in a desert having internal drainage.

plucking Glacial erosion caused as water freezes onto rock, then moves on, tearing away rock fragments.

plume A rising column of magma in the asthenosphere.

pluton A body of plutonic rock.

plutonic Igneous rock crystallized at depth.

pluvial lakes Lakes formed through abundant rainfall during ice ages.

point bar A sedimentary feature built in the stream channel on the inside of a meander or anywhere the water slows.

polar-wander curve A curve mapping past magnetic pole positions relative to a given region or continent.

polymorphs Minerals having the same chemical composition but different crystal structures.

porosity The proportion of void space (cracks, pores) in a rock.

porphyry An igneous rock with coarse crystals in a fine-grained groundmass.

potentiometric surface A theoretical surface indicating the elevation corresponding to hydrostatic pressure in a confined aquifer; analogous to the water table in an unconfined aquifer.

Precambrian The eon spanning the time from the formation of the earth to the start of the Phanerozoic.

precursor phenomena Detectable changes that occur prior to volcanic eruptions or earthquakes that might be used in prediction efforts.

Principle of Original Horizontality The concept that sedimentary rocks are generally deposited in horizontal layers, so that deviations from the horizontal reflect postdepositional disturbance.

Principle of Superposition The concept that, in an undisturbed sedimentary section, the rocks on the bottom are the oldest, with the overlying rocks progressively younger toward the top of the sequence.

proton A positively charged subatomic particle, generally found in an atomic nucleus.

P waves Compressional seismic body waves.

pyroclastics Fragments of rock and lava emitted during an explosive volcanic eruption.

pyroxenes Single-chain silicates, mostly ferromagnesian.

quartz The simplest framework silicate, with formula SiO_2.

quartzite Metamorphosed quartz-rich sandstone.

quick clay An unstable, failure-prone clay sediment, derived from glacial rock flour deposited in a marine setting, weakened by later flushing with fresh water.

radial drainage Streams radiating outward from a topographic high, such as a mountain.

radioactivity Spontaneous decay or breakdown of unstable atomic nuclei.

radiometric age A numerical date determined by the use of radioisotopes.

rain shadow A dry zone landward of coastal mountain ranges, caused by loss of moisture from air passing over mountains.

recessional moraine End moraine deposited by a retreating glacier during stationary periods.

recharge Set of processes by which groundwater is replenished.

recrystallization Process by which crystals in a rock may grow and change shape while the rock remains a solid.

rectangular drainage A fracture-controlled drainage pattern in which streams make right-angle bends.

recumbent fold A fold in which the axial plane is close to horizontal.

recurrence interval The average length of time between floods of given severity on a given stream.

refraction Deflection or change in direction of seismic body waves as they move across a boundary between two materials of different properties.

regional metamorphism Metamorphism on a large, or regional, scale; may be associated with mountain building; involves increases of both pressure and temperature.

relative dating Determining the sequence of rocks or events indicated by a particular rock section.

remote sensing Investigation or examination using light or other radiation rather than by direct contact; examples include use of aerial photography, satellite imagery, or radar.

renewable Describes a resource that can be replenished on a human time scale.

replacement The process of fossil preservation in which original material is replaced by mineral material such as silica or pyrite.

reserves That quantity of a mineral or fuel that has been found and can be exploited economically with existing technology.

reservoir rocks Rocks in which petroleum has accumulated.

resources Reserves, plus that quantity of a useful mineral or fuel known or believed to exist but not currently exploitable economically.

retention pond A basin used to hold water back from a stream temporarily after rain or a melting event, to reduce the risk of flooding.

reverse fault A dip-slip fault in which the hanging wall moves upward relative to the footwall.

rhyolite Silicic volcanic rock, the volcanic equivalent of granite.

rift valley Depression formed by grabens along the crest of a seafloor spreading ridge or on a continent.

rill erosion Soil erosion on sloping land by water forming very small channels.

ripple marks Rippled surface formed on sediment by wind or water.

rock A solid, cohesive aggregate of one or more minerals or mineral materials.

rock cycle The concept that all rocks are continually subject to change and that any rock can be transformed through appropriate geologic processes into another type of rock.

rock flour Silt-sized sediment produced by glacial abrasion.

rupture Breakage or failure under stress.

saltation A process of sediment transport in a series of short jumps along the ground or streambed.

saltwater intrusion Replacement of fresh pore water by saline water as the fresh water is depleted.

sandstone A clastic sedimentary rock made up of sand-sized particles.

scale (map) Ratio of a unit of length on the map to the corresponding actual horizontal distance represented.

scarp A steep cliff resulting from vertical displacement along a fault; also, a similar feature formed by mass movement.

schist A medium- to coarse-grained metamorphic rock displaying schistosity.

schistosity The growth of coarse, platy minerals, especially micas, in parallel planes, due to directed stress.

scientific method Means of discovering scientific principles by formulating hypotheses, making predictions from them, and testing the predictions.

seafloor spreading The process through which plates diverge and new lithosphere is created at oceanic ridges.

seamounts Volcanic hills rising a kilometer or more above the sea floor.

sediment An unconsolidated accumulation of rock, mineral grains, and organic matter that has been transported and deposited by wind, water, or ice.

sedimentary Of or relating to sediment.

seismic gap A seismically quiet section of an active fault zone, where the fault is presumed to be locked.

seismic shadow An effect of the liquid outer core, which partially deflects P waves and totally blocks S waves originating on one side of the earth from reaching the opposite side of the earth.

seismic tomography Technique for exploring earth's interior using variations in seismic-wave velocities to map lateral variations in temperature.

seismic waves The form in which energy is released during earthquakes; divided into *body waves* and *surface waves*.

sensitive clay A weak, failure-prone clay sediment similar in behavior to quick clay but derived from other materials (for example, weathered volcanic ash).

shale A clastic sedimentary rock made of clay-sized particles, having a tendency to break along parallel planes.

shearing stress Stress tending to cause different parts of an object to slide past each other across a plane.

shear strength The ability of a solid material to resist shearing stress.

sheet silicate A silicate in which tetrahedra are linked in two dimensions by shared oxygen atoms.

sheet wash Water flow over a sloping land surface; not confined to a channel.

shield A large, stable, continental region consisting of exposed Precambrian igneous and metamorphic rocks.

shield volcano A volcano with a low, flat, broad shape, formed by the buildup of many thin lava flows.

shoreline The line along which the land and water surfaces meet.

silicate A mineral containing silicon and oxygen, with or without other elements.

silicic Rich in silica (SiO_2).

sill A tabular, concordant pluton.

sinkhole A circular depression formed by ground collapse into a solution cavity.

slate A low-grade, fine-grained metamorphic rock exhibiting cleavage along parallel planes, due to alignment of clays, micas, and other sheet silicates.

slaty cleavage Rock cleavage characteristic of slate.

slide Movement of a coherent mass of rock or soil along a well-defined plane or surface.

slip face The downwind side of a dune; assumes a slope equal to the dune sediment's angle of repose.

slump A short-distance slide.

soil Surface accumulation of weathered rock and organic matter, overlying the bedrock from which it formed; generally also defined as capable of supporting plant growth.

soil moisture Water in the vadose zone.

solid solution Phenomenon of substitution of one element for another in a mineral, within some compositional limits; also, a mineral in which this occurs.

solifluction Flow in wet soil above permafrost layer in alpine terrain.

sorting Separation of minerals in a sediment by grain size; also, a measure of the extent to which this has occurred.

source The point at which a stream originates.

specific gravity The density of a mineral divided by the density of water.

spheroidal weathering Chemical weathering of a rock into a spheroidal shape, in a series of concentric layers.

spoil banks Piles or rows of waste rock and soil left after strip-mining.

spring A site where the water table intersects the ground surface so that water flows out at the surface.

spring tides Most extreme tides, which occur when sun, moon, and earth are aligned.

stage The elevation of the water surface of a stream at a given point along the channel.

stock A pluton similar to, but smaller than, a batholith.

strain Deformation resulting from stress.

stratovolcano See *composite volcano*.

streak The color of a mineral when powdered.

stream Any body of flowing water confined within a channel.

stress Force applied to an object.

striations (glacial) Parallel grooves cut in rock by rock fragments frozen into glacial ice.

strike Compass orientation of a line or plane as measured in the horizontal plane.

strike-slip fault A fault along which movement is horizontal only (parallel to strike).

strip mining Method of surface mining a shallow, areally extensive deposit by stripping off overlying rock, soil, and vegetation; most commonly used to mine shallow coal beds.

stromatolite A variously sized, mound-shaped, organic-sedimentary structure formed by algae.

subduction zone A convergent plate boundary at which a slab of oceanic lithosphere is being pushed beneath another plate (continental or oceanic) and carried down into the mantle.

submarine canyon A V-shaped canyon cut in the continental slope.

subsurface water All water below the ground surface.

sulfate A nonsilicate mineral containing sulfur and oxygen in sulfate (SO_4) groups.

sulfide A nonsilicate mineral containing sulfur but lacking oxygen.